Policy Issues in the Development of Personalized Medicine in Oncology

WORKSHOP SUMMARY

Margie Patlak and Laura Levit, *Rapporteurs*

National Cancer Policy Forum
Board on Health Care Services

INSTITUTE OF MEDICINE
OF THE NATIONAL ACADEMIES

THE NATIONAL ACADEMIES PRESS
Washington, D.C.
www.nap.edu

THE NATIONAL ACADEMIES PRESS   500 Fifth Street, N.W.   Washington, DC 20001

NOTICE: The project that is the subject of this report was approved by the Governing Board of the National Research Council, whose members are drawn from the councils of the National Academy of Sciences, the National Academy of Engineering, and the Institute of Medicine.

This study was supported by Contract Nos. HHSN261200611002C, 200-2005-13434 TO #1, and 223-01-2460 to #27, between the National Academy of Sciences and the National Cancer Institute, the Centers for Disease Control and Prevention, and the Food and Drug Administration, respectively. This study was also supported by the American Cancer Society, the American Society of Clinical Oncology, the Association of American Cancer Institutes, and C-Change. Any opinions, findings, conclusions, or recommendations expressed in this publication are those of the author(s) and do not necessarily reflect the view of the organizations or agencies that provided support for this project.

International Standard Book Number-13:   978-0-309-14575-6
International Standard Book Number-10:   0-309-14575-9

Additional copies of this report are available from the National Academies Press, 500 Fifth Street, N.W., Lockbox 285, Washington, DC 20055; (800) 624-6242 or (202) 334-3313 (in the Washington metropolitan area); Internet, http://www.nap.edu.

For more information about the Institute of Medicine, visit the IOM home page at: **www.iom.edu.**

Printed in the United States of America

The serpent has been a symbol of long life, healing, and knowledge among almost all cultures and religions since the beginning of recorded history. The serpent adopted as a logotype by the Institute of Medicine is a relief carving from ancient Greece, now held by the Staatliche Museen in Berlin.

Cover art created by Tim Cook and used with permission from the National Institutes of Health, 2004.

Suggested citation: IOM (Institute of Medicine). 2010. *Policy issues in the development of personalized medicine in oncology: Workshop summary.* Washington, DC: The National Academies Press.

"Knowing is not enough; we must apply.
Willing is not enough; we must do."
—Goethe

INSTITUTE OF MEDICINE
OF THE NATIONAL ACADEMIES

Advising the Nation. Improving Health.

THE NATIONAL ACADEMIES
Advisers to the Nation on Science, Engineering, and Medicine

The **National Academy of Sciences** is a private, nonprofit, self-perpetuating society of distinguished scholars engaged in scientific and engineering research, dedicated to the furtherance of science and technology and to their use for the general welfare. Upon the authority of the charter granted to it by the Congress in 1863, the Academy has a mandate that requires it to advise the federal government on scientific and technical matters. Dr. Ralph J. Cicerone is president of the National Academy of Sciences.

The **National Academy of Engineering** was established in 1964, under the charter of the National Academy of Sciences, as a parallel organization of outstanding engineers. It is autonomous in its administration and in the selection of its members, sharing with the National Academy of Sciences the responsibility for advising the federal government. The National Academy of Engineering also sponsors engineering programs aimed at meeting national needs, encourages education and research, and recognizes the superior achievements of engineers. Dr. Charles M. Vest is president of the National Academy of Engineering.

The **Institute of Medicine** was established in 1970 by the National Academy of Sciences to secure the services of eminent members of appropriate professions in the examination of policy matters pertaining to the health of the public. The Institute acts under the responsibility given to the National Academy of Sciences by its congressional charter to be an adviser to the federal government and, upon its own initiative, to identify issues of medical care, research, and education. Dr. Harvey V. Fineberg is president of the Institute of Medicine.

The **National Research Council** was organized by the National Academy of Sciences in 1916 to associate the broad community of science and technology with the Academy's purposes of furthering knowledge and advising the federal government. Functioning in accordance with general policies determined by the Academy, the Council has become the principal operating agency of both the National Academy of Sciences and the National Academy of Engineering in providing services to the government, the public, and the scientific and engineering communities. The Council is administered jointly by both Academies and the Institute of Medicine. Dr. Ralph J. Cicerone and Dr. Charles M. Vest are chair and vice chair, respectively, of the National Research Council.

www.national-academies.org

WORKSHOP PLANNING COMMITTEE[1]

ROY HERBST (*Cochair*), Professor and Chief, Section on Thoracic Medical Oncology, Department of Thoracic/Head and Neck Medical Oncology, M.D. Anderson Cancer Center, Houston, TX

DAVID PARKINSON (*Cochair*), President and Chief Executive Officer, Nodality, Inc., San Francisco, CA

FRED APPELBAUM, Director, Clinical Research Division and Head, Division of Medical Oncology, Fred Hutchinson Cancer Research Center, Seattle, WA

PETER BACH, Associate Attending Physician, Memorial Sloan-Kettering Cancer Center, New York, NY

ROBERT ERWIN, President, Marti Nelson Cancer Foundation, Davis, CA

STEPHEN FRIEND, President, Chief Executive Officer, and Cofounder, Sage Bionetworks, Seattle, WA

STEVEN GUTMAN, Professor of Pathology, University of Central Florida, Orlando, FL

GAIL JAVITT, Law and Policy Director, Genetics and Public Policy Center, Johns Hopkins University, Washington, DC

SAMIR KHLEIF, Senior Investigator and Chief of Cancer Vaccine Section, National Cancer Institute, Bethesda, MD

Study Staff

LAURA LEVIT, Study Director
CASSANDRA L. CACACE, Research Assistant
MICHAEL PARK, Senior Program Assistant
ASHLEY McWILLIAMS, Senior Program Assistant
PATRICK BURKE, Financial Associate
SHARYL J. NASS, Director, National Cancer Policy Forum
ROGER HERDMAN, Director, Board on Health Care Services
SHARON B. MURPHY, Scholar in Residence

---

[1] Institute of Medicine planning committees are solely responsible for organizing the workshop, identifying topics, and choosing speakers. The responsibility for the published workshop summary rests with the workshop rapporteurs and the institution.

NATIONAL CANCER POLICY FORUM[1]

HAROLD MOSES (*Chair*), Director Emeritus, Vanderbilt-Ingram
Cancer Center, Nashville, TN
FRED APPELBAUM, Director, Clinical Research Division, Fred
Hutchinson Cancer Research Center, Seattle, WA
PETER B. BACH, Associate Attending Physician, Memorial Sloan-
Kettering Cancer Center, New York, NY
EDWARD BENZ, JR., President, Dana-Farber Cancer Institute and
Director, Harvard Cancer Center, Harvard School of Medicine,
Boston, MA
THOMAS G. BURISH, Past Chair, American Cancer Society Board of
Directors and Provost, Notre Dame University, South Bend, IN
MICHAELE CHAMBLEE CHRISTIAN, Retired, Division of Cancer
Treatment and Diagnosis, National Cancer Institute, Bethsda, MD
ROBERT ERWIN, President, Marti Nelson Cancer Foundation,
Davis, CA
BETTY R. FERRELL, Research Scientist, City of Hope National
Medical Center, Duarte, CA
JOSEPH F. FRAUMENI, JR., Director, Division of Cancer
Epidemiology and Genetics, National Cancer Institute,
Bethesda, MD
PATRICIA A. GANZ, Professor, University of California, Los Angeles,
Schools of Medicine & Public Health, Division of Cancer Prevention
& Control Research, Jonsson Comprehensive Cancer Center, Los
Angeles, CA
ROBERT R. GERMAN, Associate Director for Science (Acting),
Division of Cancer Prevention and Control, Centers for Disease
Control and Prevention, Atlanta, GA
ROY S. HERBST, Chief, Thoracic/Head & Neck, Medical Oncology,
M.D. Anderson Cancer Center, Houston, TX
THOMAS J. KEAN, Executive Director, C-Change, Washington, DC
JOHN MENDELSOHN, President, M.D. Anderson Cancer Center,
Houston, TX

---

[1] IOM forums and roundtables do not issue, review, or approve individual documents.
The responsibility for the published workshop summary rests with the workshop rapporteurs
and the institution.

JOHN E. NIEDERHUBER, Director, National Cancer Institute, Bethesda, MD

DAVID R. PARKINSON, President and Chief Executive Officer, Nodality, Inc., San Francisco, CA

SCOTT RAMSEY, Full Member, Cancer Prevention Program, Fred Hutchinson Cancer Research Center, Seattle, WA

JOHN WAGNER, Executive Director, Clinical Pharmacology, Merck and Company, Inc., Whitehouse Station, NJ

JANET WOODCOCK, Deputy Commissioner and Chief Medical Officer, Food and Drug Administration, Rockville, MD

National Cancer Policy Forum Staff

SHARYL NASS, Director, National Cancer Policy Forum
LAURA LEVIT, Program Officer
CHRISTINE MICHEEL, Program Officer
ERIN BALOGH, Research Associate
ASHLEY McWILLIAMS, Senior Program Assistant
MICHAEL PARK, Senior Program Assistant
PATRICK BURKE, Financial Associate
SHARON B. MURPHY, Scholar in Residence
ROGER HERDMAN, Director, Board on Health Care Services

Reviewers

This report has been reviewed in draft form by individuals chosen for their diverse perspectives and technical expertise, in accordance with procedures approved by the National Research Council's Report Review Committee. The purpose of this independent review is to provide candid and critical comments that will assist the institution in making its published report as sound as possible and to ensure that the report meets institutional standards for objectivity, evidence, and responsiveness to the study charge. The review comments and draft manuscript remain confidential to protect the integrity of the process. We wish to thank the following individuals for their review of this report:

ELI ADASHI, Professor of Medical Sciences, Brown University, Providence, RI

STEVEN GUTMAN, Professor of Pathology, University of Central Florida, Orlando, FL

GAIL JAVITT, Law and Policy Director, Genetics and Public Policy Center, Johns Hopkins University, Washington, DC

MUIN KHOURY, Director, Office of Public Health Genomics, Centers for Disease Control and Prevention, Atlanta, GA

Although the reviewers listed above have provided many constructive comments and suggestions, they were not asked to endorse the final draft of the report before its release. The review of this report was overseen by **Melvin Worth**. Appointed by the Institute of Medicine, he was responsible for making certain that an independent examination of this report was carried out in accordance with institutional procedures and that all review comments were carefully considered. Responsibility for the final content of this report rests entirely with the authors and the institution.

Contents

Introduction

Personalized cancer medicine is defined as medical care based on the particular biological characteristics of the disease process in individual patients. By using genomics and proteomics, individuals can be classified into subpopulations based on their susceptibility to a particular disease or response to a specific treatment. They may then be given preventive or therapeutic interventions that will be most effective given their particular characteristics.

In oncology, personalized medicine has the potential to be especially influential in patient treatment because of the complexity and heterogeneity of each form of cancer. However, the current classifications of cancer are not as useful as they need to be for making treatment decisions; current cancer classification evolved from morphology and may be misleading because it does not take into account abnormalities at the molecular level. As a result, treatment needs to evolve toward a focus on targeted treatments based on individual characterizations of the disease.

Although this concept has great promise, a number of policy issues must be clarified and resolved before personalized medicine can reach its full potential. These include technological, regulatory, and reimbursement hurdles. To explore those challenges, the National Cancer Policy Forum held a workshop, "Policy Issues in the Development of Personalized Medicine in Oncology," in Washington, DC, on June 8 and 9, 2009. At this workshop experts gave presentations and commentary on the following areas:

- The current state of the art of personalized medicine technology, including obstacles to its development and use by clinicians and patients.
- The current approaches to test validation, including analytic validity, clinical validity, and clinical utility.
- The regulation of personalized medicine technologies, including the approaches' shortcomings.
- Reimbursement hurdles that can hamper both the development and use of personalized medicine technologies.
- Potential solutions to the technological, regulatory, and reimbursement obstacles to personalized medicine.

This document is a summary of the conference proceedings, which will be used by an Institute of Medicine (IOM) committee to develop consensus-based recommendations for moving the field of personalized cancer medicine forward. The views expressed in this summary are those of the speakers and discussants, as attributed to them, and are not the consensus views of the participants of the workshop or of the members of the National Cancer Policy Forum.

Personalized Cancer Medicine Technology

Several speakers illustrated both the accomplishments of personalized cancer medicine and the challenges that remain ahead, using examples in the treatment of leukemia, breast, colon, and lung cancer. These speakers discussed a number of tests that predict patient response to specific cancer treatments, including tests for the following:

- HER2, which predicts a patient with breast cancer's response to Herceptin.
- Estrogen receptors, which predict a patient with breast cancer's response to tamoxifen and aromatase inhibitors.
- Mutations in the epidermal growth factor receptor (EGFR), which are predictive of a patient with lung cancer's response to drugs such as gefitinib or erlotinib. The mutations also predict response when drugs that target EGFR are used in combination with other cytotoxic chemotherapies.
- Mutations in the KRAS protein that play an important role in EGFR signaling, and predict an individual's response to colon cancer drugs that act on this receptor, such as cetuximab.
- Mutations in the tyrosine kinase receptor FLT3, which confer resistance to drugs that target the receptor in patients with leukemia.
- Gene expression variations in tumors that predict breast cancer recurrence (Oncotype DX, MammaPrint).

- Drug metabolism genetic variants that predict adverse reactions to the cancer drug irinotecan.

Many of the tests that are predictive of a therapeutic response (here-inafter, in this report, "predictive tests") have regulatory approval and are the standard of care for certain cancer treatments. The breast cancer drug Herceptin, as well as the tests that indicate patients likely to respond to it, has been on the market since 1998 and has been used to treat half a million patients (Roche, 2008). More than 100,000 Oncotype Dx tests, a gene expression test that predicts a patient's benefit from chemotherapy as well as breast cancer recurrence, have also been used to determine treatment planning since the test came on the market in 2004 (Genomic Health, 2009). About half of all estrogen-positive breast tumors in the United States are being evaluated with this preditive test, estimated Dr. Steven Shak of Genomic Health, the test's developer. In addition, the UGT1A1 molecular assay has Food and Drug Administration (FDA) clearance for patients with colorectal cancer who are considering taking Camptosar (irinotecan), and tests for KRAS are approved by the European Medicines Agency (EMEA) to predict patients' response to panitumumab and cetuximab therapy in colorectal cancer.[1] Phase III clinical trials have recently confirmed the predictive value of EGFR mutations for response to gefitinib (Iressa) and erlotinib (Tarveva), leading the EMEA to announce its approval of gefi-tinib as a treatment for lung tumors that have activating EGFR mutations (AstraZeneca, 2009).

Predictive tests can be useful in health care because they often calculate an individual's response to treatment better than other clinical indicators, said Dr. Bruce E. Johnson of the Dana-Farber Cancer Institute. For example, non-smoking women with a particular type of lung cancer are more likely to respond to erlotinib or gefitinib than other patients with lung cancer. Patients meeting these clinical characteristics have a median progression-free survival (PFS) of about 6 months, compared to a median PFS of less than 3 months in individuals without these clinical features. However, median PFS was nearly 15 months in individuals with EGFR mutations that predict response to erlotinib, versus only about 2 months in individuals without these mutations (see Figures 1a and 1b). Dr. Johnson and Dr. Rafael Amado of GlaxoSmithKline noted the importance of showing, with appropriately

---

[1] A similar decision was made by the FDA shortly after the workshop.

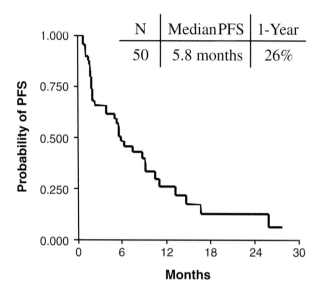

N	Median PFS	1-Year
50	5.8 months	26%

FIGURE 1a Clinically enriched patients. Non-smoking women with a particular type of lung cancer are more likely to respond to erlotinib or gefitinib than other patients with lung cancer. Patients meeting these clinical characteristics have a median progression-free survival (PFS) of about 6 months.
SOURCES: Johnson presentation (June 8, 2009); Bruce Johnson and David Jackman, Dana-Farber Cancer Institute.

designed clinical trials, that a predictive test truly predicts response to treatment, rather than indicating a prognosis independent of treatment.

A potential benefit of predictive tests is that they limit the number of individuals who will have an adverse or ineffective response to a therapeutic treatment. For example, the use of Oncotype DX reduces overall chemotherapy use by at least 20 percent (Shak, 2009). "There are a number of patients who are no longer receiving therapy uselessly, and there has been a lot of money saved," said Dr. Amado. However, Dr. Mark Ratain of the University of Chicago Hospitals said that "the more we learn, the more we know we don't know." Deciphering the clinical implications of predictive tests can be challenging, even when they assess the function of just one key protein. Genetic assessments are likely to become more complex in the future. As a result, it will become necessary for researchers to develop multiple predictive tests that indicate the function of many, if not all, the nodes on those pathways that play crucial roles in the development or progression

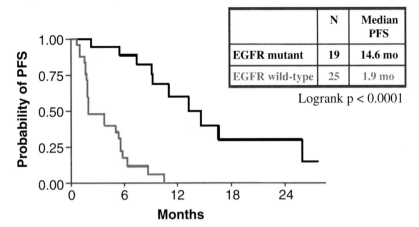

	N	Median PFS
EGFR mutant	19	14.6 mo
EGFR wild-type	25	1.9 mo

Logrank p < 0.0001

FIGURE 1b Genomically defined patients. Median progression-free survival (PFS) was nearly 15 months in individuals with lung cancer and epidermal growth factor receptor (EGFR) mutations that predict response to erlotinib, versus only about 2 months in individuals without these mutations.
SOURCES: Johnson presentation (June 8, 2009); Bruce Johnson and David Jackman, Dana-Farber Cancer Institute.

of various cancers. Dr. Stephen Friend of Sage Bionetworks suggested that because of redundant backup pathways and feedback loops, scientists need to model and consider entire pathway networks when developing predictive tests.

DECIPHERING THE CLINICAL IMPLICATIONS

Dr. Donald Small of the Sidney Kimmel Comprehensive Cancer Center illustrated some of the difficulties of making treatment decisions based on the results of predictive tests. For example, treatment decisions for patients with acute myelogenous leukemia (AML) are often based on the results of tests for mutations on the tyrosine kinase receptor FLT3. This receptor plays a role in stimulating the proliferation of blood stem cells and dendritic cells of the immune system. Researchers have discovered a number of mutations on this gene, as well as in the DNA stretch that controls its activation, which affect the responsiveness of patients with AML to FLT3 inhibitor drugs. However, the mere presence of specific mutations does not determine responsiveness to anti-FLT3 treatment. Rather, the ratio of the mutant gene to the wild-type allele predicts responsiveness (Smith et al., 2004). Patients with the lowest ratio of the mutant gene to the wild-type allele have the best clinical prognosis (Figure 2) (Meshinchi et al., 2006). Complicating the clinical decision making, however, is evidence that patients with FLT3 mutations who receive a bone marrow transplant have similar outcomes to those patients without mutations. As a result, some clinicians are inclined to treat patients with AML with a bone marrow transplant, rather than treating them with a FLT3 inhibitor.

Another example of how the development of predictive tests may outpace the clinical understanding of these tests is in the use of Oncotype DX. A high recurrence score from an Oncotype DX test indicates those women with estrogen receptor-positive (ER-positive), node-negative breast cancer who are at high risk for relapse and most likely to benefit from adjuvant chemotherapy. A low recurrence score indicates women who should only receive hormonal therapy (Paik et al., 2006). However, the test does not provide useful information on how women whose scores are in the middle range should be treated. The clinical study, TailoRx, is currently assessing the predictive value of these mid-range scores (NCI, 2009b), but in the meantime clinicians are unsure what the best treatment is for women with these intermediate scores.

"I recently tried to help a woman who had been diagnosed with a small ER-positive breast cancer with no lymph node involvement," said Amy Bonoff of the National Breast Cancer Coalition. "But she had a gene assay test that showed she was in the high middle range for risk of recurrence. What should she do? No one has the answer to that. She now has a piece of information that will keep her awake at night, and she really can't make medical decisions" based on it. Ms. Bonoff stressed that "for a biomarker to be clinically meaningful it must improve patient outcomes in a meaningful way, and predict disease outcome in the absence of treatment or guide the use of therapy targeted to the marker." Dr. Richard Schilsky of the University of Chicago and the Cancer and Leukemia Group B (CALGB), added, "Biomarker development needs to start off by defining the intended use of the test. If we can't define what it's going to be used for, why develop it?" However, Dr. Shak noted that personalized medicine requires the integration of other prognostic factors, such as tumor size and grade, with genetic factors. "These factors all need to be taken into account. Oncotype DX is not a recipe," he said.

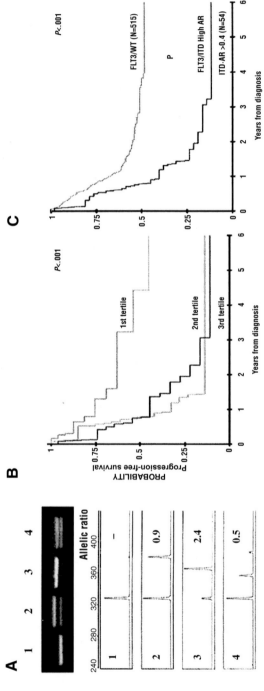

FIGURE 2 Allelic ratio (mutant to wild-type FLT3 allele) affects the prognostic significance of FLT3/ITD mutations. (A) Example of ITD-AR determination by Genescan analysis. The top panel is the agarose gel resolution of PCR product from a normal marrow (lane 1) and specimens from 3 patients with FLT3/ITD (lanes 2–4). The lower panels show the result of the Genescan analysis and ITD-AR determination. (B) Actuarial progression-free survival (PFS) from study entry for patients with FLT3/ITD based on allelic ratio by tertiles. (C) Actuarial PFS from study entry for patients with high ITD-AR (ITD-AR > 0.4) compared with those with FLT3/WT. Patients were from a Children's Oncology Group acute myelogenous leukemia trial.

SOURCES: Small presentation (June 8, 2009); Meshinchi et al. (2006). This research was originally published in *Blood.* Meshinchi, S., T. A. Alorzo, D. L. Stirewalt, M. Zwaan, M. Zimmerman, D. Reinhardt, G. J. Kaspers, N. A. Heerema, R. Gerbing, B. J. Lange, and J. P. Radich. Clinical implications of FLT3 mutations in pediatric AML. 2006; Vol 108(12):3654–3661. © the American Society of Hematology.

INCREASING COMPLEXITY OF PREDICTIVE TESTS

The use of the KRAS test in patients with colorectal cancer demonstrates the need for more complex predictive testing, and a better understanding of how predictive tests work. It is standard practice to only treat colorectal cancer patients with EGFR-targeting drugs if they have the KRAS genetic profile that is likely to render them responsive to such treatment. The use of KRAS genotyping results in a near doubling of response rate and progression-free survival of patients with colorectal cancer treated with these medicines, compared to an unselected patient population, Dr. Amado said (Jonker et al., 2007). However, these are marginal results because the response rate is still only about 20 percent in patients with the correct KRAS genetic profile. "Clearly there's more beyond KRAS," he said.

KRAS is a node on one of two pathways thought to be essential for EGFR signaling. A key node on the other pathway is P13K (Figure 3) (Scaltriti and Baselga, 2006). Recent data reveal that mutations in KRAS do not affect an individual's sensitivity to anti-EGFR treatments. Instead, mutations in an effector protein downstream from KRAS, called B-Raf, predicts response to anti-EGFR treatment independent of KRAS mutations (Di Nicolantonio et al., 2008). About 10 percent of colorectal patients

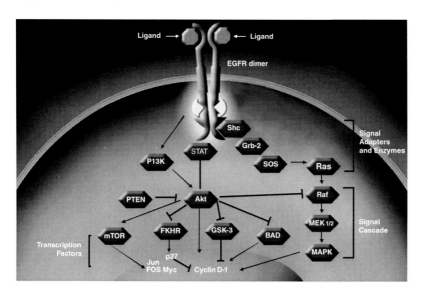

FIGURE 3 EGFR signal transduction.
SOURCE: Amado presentation (June 8, 2009).

have B-Raf mutations, 30 percent have wild-type KRAS with B-Raf, and 60 percent have B-Raf mutations and wild-type KRAS. Mutations in either of these two genes predicts lack of response to cetuximab (Di Nicolantonio et al., 2008). Preliminary data also suggest that levels of expression of certain ligand proteins (AREG or EREG) predict responsiveness to anti-EGFR treatment in colorectal cancer patients independent of KRAS status. One study found that a "combimarker" (i.e., detecting KRAS mutations and expression levels of these ligand proteins) could select a population with an overall survival ratio of .43, compared to a ratio of .7 if no markers are used to select patients (Jonker et al., 2009). "What these data are suggesting is that it's not really about a single node in the pathway, but rather about the pathway itself," said Dr. Amado. "If we're looking at genes in isolation, we may make incremental movement forward, but ideally in the future, we should have techniques that are really looking down that pathway that's activated for individual tumors. Hopefully our predictive test capability will evolve in that direction."

Aiding that evolution are genomics technologies, which give researchers the opportunity to assay large sets of genetic markers simultaneously to determine the "genetic signatures" that correlate with prognosis and/or responsiveness to treatment. Dr. Friend described several predictive tests that examine large sets of genetic markers that use this technology, including an FDA-cleared, 70-gene expression test called MammaPrint, which predicts women likely to experience a recurrence of their breast cancer, and the Onco-type DX test (Paik et al., 2004; van't Veer et al., 2002). He pointed out that genetic signatures can distinguish between tumors that are ER positive and negative and those that are HER2 positive and negative, suggesting that the signatures correlate well with the underlying biology of the tumors.

Dr. Friend also described research that used cells in culture or tumor cells in mice to discern the groups of genes that are upregulated or down-regulated by RAS or RAS inhibitors (Bild et al., 2006; Blum et al., 2007; Sweet-Cordero et al., 2005). This work revealed that whole sets of genes can act like switches—turn on or off—in response to certain drugs or proteins. He suggested that research should focus on identifying genetic signatures in patients' tumors that indicate whether their cancer-promoting pathways are likely to be blocked by treatment. For example, Dr. Friend and his colleagues developed a 147-gene signature that assesses the RAS pathway as a whole, and identifies, with greater than 90 percent sensitivity, KRAS-mutant lung tumors and cancer cell lines (Friend, 2009).

Interestingly, there is an overlap of only one gene in the MammaPrint and Oncotype DX genetic signature, and an overlap of 14 genes in the Merck RAS genetic signature and another RAS signature (Friend, 2009). Dr. Friend stressed the importance of ascertaining why there is not more overlap between the various genetic signatures that predict the same outcomes, and noted that as more signatures are developed, it will be difficult to decide which ones are the best ones to put into practice.

Dr. Friend also called for a better understanding of the pathways being tested. More insight is needed into the overarching causal mechanisms that are driving the cancer, including an awareness of redundant feedback loops he called networks, which become active when the pathways are blocked. "Not only do you have to have the markers, but you also have to understand the pathway and the network that's sitting behind it," he said. "If you look at the data that are coming, the data are miniscule compared to what's going to happen in the next 5 or 10 years. We'll have the ability to have a DNA sequence across the entire tumor on most patients and then look also at expression profiling, because you can do it at the same time." Dr. Ratain concurred, stating that "our current strategy in pharmacogenomics is to collect DNA samples in conjunction with large clinical trials and to perform genome-wide typing to identify candidates associated with both toxicity and efficacy. Then we can conduct replication studies using samples from other similar studies, and perform mechanistic studies to confirm function." A recent study used such a strategy to show a genomic basis for an adverse reaction to statin treatment (statin myopathy) (Search Collaborative Group et al., 2008). "This shows the power of genome-wide association for discovery of functional variants," Dr. Ratain said.

Dr. Friend stressed the need to integrate different types of genomic information, and using Bayesian approaches, build up probabilistic causal models of disease that go beyond just looking at markers on a pathway. He and his colleagues used such an approach to build a model of obesity that indicated that nine genes were key players in the disorder (Schadt et al., 2005). A validation study then showed that eight of those nine genes modulate obesity when they are overexpressed, altered, or knocked out (Yang et al., 2009). "We can now build predictive, causal networks," he said. "When you go to a tumor state, instead of ranking genes that are altered, we think it's much better to actually look at the networks that are broken and reassociate them" (Figure 4).

However, such assessments require collaboration on a large scale. "No one company or institution should or could build these probabilistic causal maps," Dr. Friend said. "It won't work if we work in fiefdoms. We need to

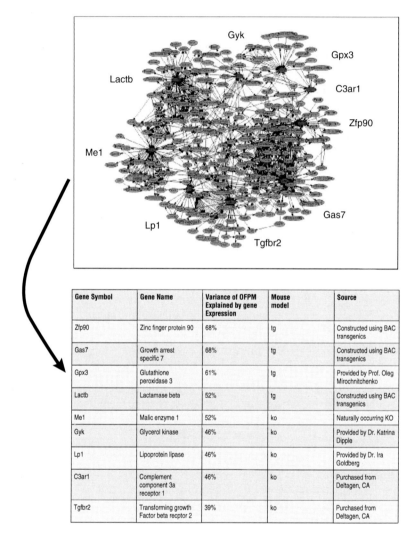

FIGURE 4 Networks facilitate direct identification of genes that are causal for disease (obesity).

SOURCES: Friend presentation (June 8, 2009) and Schadt et al. (2005); Yang et al. (2009). Reprinted by permission from Macmillan Publishers Ltd: Nature Genetics (Yang, X., J. L. Deignan, H. Qi, J. Zhu, S. Qian, J. Zhong, G. Torosyan, S. Majid, B. Falkard, R. R. Kleinhanz, J. Karlsson, L. W. Castellani, S. Mumick, K. Wang, T. Xie, M. Coon, C. Zhang, D. Estrada-Smith, C. R. Farber, S. S. Wang, A. van Nas, A. Ghazalpour, B. Zhang, D. J. MacNeil, J. R. Lamb, K. M. Dipple, M. L. Reitman, M. Mehrabian, P. Y. Lum, E. E. Schadt, A. J. Lusis, and T. A. Drake. 2009. Validation of candidate causal genes for obesity that affect shared metabolic pathways and networks. *Nature Genetics* 41(4):415–423.), Copyright (2009).

create a commons where scientists can combine their datasets with others to build network models. Chemists and physicists have used structures and models of what they work on for decades. The irony is that doctors don't. They really don't have molecular, physiologic models of disease, but rather little pathway maps that have worked as examples."

Dr. Friend recently formed a nonprofit organization called Sage Bionetworks. This organization will provide a commons for the creation of disease models based on the assembly of coherent biomedical data into probabilistic and integrative bionetworks models (Friend, 2009). These models evolve via modifications made by contributor scientists. The ultimate mission of Sage is to accelerate the elimination of human diseases.

Dr. Robert Mass of Genentech, Inc., agreed on the importance of going beyond gene expression data to understand the underlying tumor biology, but noted that even with that understanding, developing the appropriate predictive tests can be difficult. For example, examination of the HER2 tumor-promoting pathway led researchers at Genentech to discover that tumors responsive to Herceptin appeared to have dimerization of HER2, with either HER1 or HER3 (Mass, 2009). However, detecting HER2 dimerization in clinical samples is difficult to do because it requires detecting phosphorylated HER2 or activated HER2—modified forms of the proteins that are short-lived and difficult to detect in fresh tissue, and virtually impossible to reliably detect in formalin-fixed, paraffin-embedded tissue, according to Dr. Mass. As a result, researchers had to detect downstream surrogate markers, such as low levels of HER3 in ovarian cancer patients, as measured by quantitative reverse transcriptase polymerase chain reaction (PCR), and HER2 amplification in breast cancer patients. "It's going to be complicated because we may be using different markers for different groups of patients, which is a challenge to a drug developer," Dr. Mass said.

Adding to the complexity of developing personalized cancer medicine is individual variability in how much of a given drug reaches its target, Dr. Small pointed out. He noted that typically, the assays to test the effectiveness of drugs that target tyrosine kinase receptors, such as FLT3, are done in the absence of fetal calf serum or similar compounds that mimic the effects that bloodstream products have on the binding of a drug on its target. Human plasma has numerous proteins that can bind to drugs. A recent study indicated that binding can change the concentration of drugs in the bloodstream from the nanomolar range to the micromolar range, he said (Levis et al., 2006) (Figure 5). Different patients show different bind-

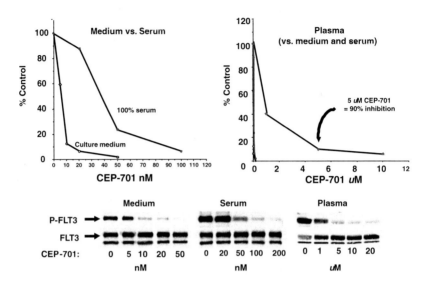

FIGURE 5 Drug binding: Inhibition of FLT3 autophosphorylation by CEP-701. SOURCES: Small presentation (June 8, 2009) and Levis et al., 2006. This work was originally published in *Blood*. Levis, M., P. Brown, B. D. Smith, A. Stine, R. Pham, R. Stone, D. DeAngelo, I. Galinsky, F. Giles, E. Estey, H. Kantarjian, P. Cohen, Y. Wang, J. Roesel, J. E. Karp, and D. Small. Plasma inhibitory activity (PIA): a pharmacodynamic assay reveals insights into the basis for cytotoxic response to FLT3 inhibitors. 2006; Vol 108(10):3477–3483. © the American Society of Hematology.

ing to FTL3 inhibitors, as determined by assays with FLT3 inhibition using patient serum. In leukemia cell lines, a drug could inhibit 80–90 percent of FLT3 receptor activity in the presence of some patients' serum, but only achieve 60–70 percent inhibition in the presence of serum from other patients. In addition, this study found that clinical response to these drugs correlated with the degree of inhibition achieved in the assays (Smith et al., 2004). "Shouldn't we be individualizing drug dosing to attain sufficient inhibition in all patients?" Dr. Small asked. "This is something that hasn't really been occurring in typical tyrosine kinase inhibitor trials."

Non-genetic sources of variability also need to be considered, Dr. Ratain pointed out. These include dose and schedule, disease severity, concomitant conditions and use of other drugs, liver and kidney function, and age. Because many new cancer treatments are oral drugs, the effect of diet on their action needs to be considered, he added. "Although I spend most of my life thinking about pharmacogenomics, particularly germ line, it all goes to naught if we don't also consider these non-genetic issues," Dr. Ratain said. Dr. Fred Appelbaum of Fred Hutchinson Cancer Research Center concurred, saying, "In oncology, so many of our patients are elderly and have a litany of comorbidities that hugely affect their tolerance to drugs and their toxicities. It's easier to look at genes and profiling. It's very hard to get all the data necessary to list all the comorbidities that will influence toxicities."

TEST VALIDATION

The characteristics of a reliable test is analytic validity (accuracy in detecting the specific entity it was designed to detect) and clinical validity (accuracy for a specific clinical purpose, e.g., predicting response to treatment). Predictive tests also should be useful in clinical decision making and in improving patient outcomes (clinical utility).

Determining the analytical validity of a predictive test is a long and arduous process, Dr. Shak said. "Just as the development of a drug cannot be achieved by performing a single study, the same thing is true with regard to the development of a predictive test and its validation." Analytic validation requires showing assay performance, standardization and analytic performance, and whether the assay performs the same under different formats and conditions. To assess analytic validity, researchers must take into account variability in sample preparation. For example, in the real-world clinical setting, there can be variability in the time from when tumor tissue is harvested in an operating suite and is placed in formalin, as well as in the time a tissue sample remains in formalin. An assay has to perform consistently under all variations in sample preparation.

The development process for a predictive test also has to be standardized and reproducible. "Typically it takes us between 6 to 12 months to look at reproducibility, and to ensure that every aspect of the assay is going to be performed properly, and all the reagents are appropriately qualified and the specifications are set. One needs to be patient in that regard—these are critically important steps that can't be avoided," said Dr. Shak. His laboratory had to specify more than 150 standard operating procedures for its 5-step Oncotype DX test.

Determining the clinical validity and utility of a predictive test can also be time consuming and challenging. These qualifications require showing that the assay is "fit for purpose," and ultimately provides some patient benefit. Typically, a retrospective/prospective study is done to clinically validate a predictive test and show its clinical utility, Dr. Schilsky explained.

Exploratory or correlative analyses are done on clinically annotated specimens that were collected prospectively. The assay methods are applied retrospectively after the clinical outcomes of the trial are known. Although prospective clinical trials are viewed as the gold standard for determining clinical utility, such retrospective/prospective trials suffice, as long as they are done in a rigorous manner (i.e., a different dataset is used for clinical utility than was used for validation, and the analyses are prespecified, robust, and show a large treatment effect), Dr. Amado said. A biologically plausible effect gives further support for the clinical utility, and may preclude the need for a prospective study, he added. Dr. Daniel Hayes of the University of Michigan Comprehensive Cancer Center concurred, saying, "If you're going to use archived samples, you have to be as rigorous as if it was a prospective trial. You have to have a prospectively written protocol, and put down the statistical power you think you're going to get. And you need more validated datasets if you're using archive samples than you would for a prospective clinical trial."

Dr. Friend cautioned that sometimes the dataset originally collected—and on which the retrospective/prospective analysis is done to show a biomarker's clinical validity or utility—may have a skewed population or bias. He suggested making sure that such biomarker studies apply to a broad population. Dr. Ratain added that researchers and clinicians should be careful about overinterpreting nonreplicated findings. "Retrospective is fine as long as it's well replicated. All too often you see findings presented at prestigious meetings that really are not well replicated." Dr. Ratain also noted that randomized trials are often "a missing metric" in the assessment of predictive tests.

Risa Stack of Kleiner Perkins Caulfield and Byers stressed that the ability to use archived samples is key to innovation in personalized medicine. Traditionally, she said, such use of archived samples has not been allowed in the FDA approval process. Without this avenue of study, companies have to do prospective studies that may take as long as 10 years to complete. By that time, the therapies for which the predictive tests were developed may no longer be relevant. However, Dr. Mansfield of the FDA pointed out that the FDA has always allowed archive samples when it is appropriate to use them, and offers a guidance document about using leftover samples that are deidentified.

Dr. Shak pointed out that it can be statistically challenging to determine the clinical validity and utility of predictive tests that use genomic or genetic microarray technology. The multiple analyses done simultaneously with these predictive tests increase the likelihood that an initial association detected as statistically significant will ultimately end up being an artifact. "The good news about looking at thousands of genes is the fact that you'll always see positive results," he said. "One of the obstacles of this field is human nature—when one sees a little bit of results in 70 patients, it's really easy to get excited and feel you're only 10 yards away from having the next best test. We need discipline and very close interaction with our statistical colleagues—both the clinical biostatisticians and the non-clinical biostatisticians—so you can identify artifacts and show reproducibility, outliers, and linearity," Dr. Shak said. For example, recent evidence reviews and recommendations by the EGAPP working group suggests there is insufficient clinical utility for several predictive tests that are currently the standard of care, and that more studies are needed (EGAPP Working Group, 2009a, 2009b).

To truly confirm initial findings and clinically validate a biomarker, researchers often have to conduct studies using large number of patient tumor samples. Several speakers noted the difficulties in acquiring sufficient numbers of tumor samples. Dr. Shak said the clinical validation study done on the Oncotype DX test would have been impossible if the National Surgical Adjuvant Breast and Bowel Project, a clinical trials cooperative group, had not preserved tissue samples it collected in the 1990s to establish the benefit of chemotherapy in women with breast cancer (Paik et al., 2004). "There should be funding that would allow us to be able to collect and save tissue blocks so we can learn from our studies," said Dr. Shak. Dr. Schilsky pointed out that the quality and variability in the biospecimens collected at various sites participating in clinical trials necessary to validate predictive tests can also be problematic. Dr. Mass called for having more repositories of frozen tumor tissue that is properly collected.

An alternative method to retrospective/prospective trials for validating a biomarker is to conduct prospective biomarker-drug codevelopment studies, in which patients are identified as biomarker positive or biomarker negative, and both groups are randomized to receive the new treatment versus standard treatment. However, accruing the large number of patients needed to validate a biomarker in this manner is a major hurdle, especially when the expected outcome is minimal, and the treatment being tested with the biomarker is already available clinically, Dr. Schilsky noted. "It's far easier to just give the treatment to the patients," he said, adding that the numbers of patients required for a biomarker validation study often far exceed the number of patients needed to assess the clinical efficacy of a drug. Dr. Mass added that "it's almost impossible to do prospective validation unless you go to some part of a developing country where no access to these drugs is available, but there are ethical challenges with doing that. These prospective validation studies are just not achievable."

Ms. Bonoff and another participant suggested tapping the advocacy community to foster more patient outreach and education on biomarkers, with the intent of encouraging more patients to participate in clinical validation trials on biomarkers. She suggested using a strategy similar to that used by Dr. Susan Love, who used the Internet to create a "million-person army" of women with breast cancer; participating women are notified of clinical studies on breast cancer, including clinical trials on breast cancer drugs (Love/Avon Army of Women, 2009). Many of these women volunteer for such trials. "We need to figure out a way to get patients themselves to say, 'I want these assays. I know they're not sound yet, and I want to help build them,' " said an unidentified participant. Dr. Debra Leonard of the Weill Cornell Medical College suggested capturing data from the medical practices of early users of predictive tests. These data could be used to analyze the clinical value of those tests, perhaps with the aid of electronic medical records.

The low level of funding for validating biomarkers has also hampered their development, several speakers asserted. Federal grants and other incentives traditionally are geared toward individual accomplishments, but the translational research needed to further personalized medicine is a collaborative process, said Dr. Shak. "The biggest policy issue to me is how we can better align all of our incentives across the board to get us working together as a team in order to deliver on the promise of personalized medicine," he said.

Dr. Schilsky raised the need for commercial partners in biomarker validation studies. Dr. Ratain said his experience was that corporate entities were uninterested in supporting his pharmacogenetic research on the metabolism of irinotecan, which led to tests that predict adverse reactions to the drug. Instead, he relied on the National Cancer Institute (NCI) for funding. Only when the FDA changed the drug label of irinotecan to include information that linked a specific genetic variant with a heightened risk of an adverse reaction to the drug did corporations show an interest in developing predictive tests for the variant, he said. The reluctance of drug companies to support the development of predictive tests is a major impediment to the transfer of this technology. "There is a lack of a corporate entity that has the financial wherewithal to really develop these tests," he said.

Dr. Schilsky added that academic collaborations with industry partners to conduct these trials often results in legal tussles over who owns the data or specimens collected, and other intellectual property right issues. "We can spend years in negotiation over these types of issues," he said.

Patent claims on predictive tests also can impede innovation if one has to acquire numerous patent licenses to develop a multigene test, and there are competing patent licenses on different sets of genes, Dr. Ratain pointed out. Dr. Mass commented that a use patent on Oncotype DX should prevent people from using the same 21 genes in the assay in the same way, but should not prevent investigators from striving to improve such assays using some of those genes or using the same genes, but in a different way or for a different purpose.

Another factor that can hamper biomarker development and validation is the requirement that academic laboratories conducting predictive tests must achieve the Clinical Laboratory Improvement Amendments of 1988 (CLIA)[2] certification, Dr. Schilsky said. "This is a huge issue in making the transition from moving an assay from an academic research lab into a more clinically informative setting," he said. "We've had CALGB trials we have been doing for which we've had to find alternative laboratories in the middle of the trial because all of a sudden this stringency about using CLIA-certified laboratories has increased, and we've had to say to a research lab that's been doing an assay for years, 'You can't do this assay anymore because you're not CLIA certified.' It's a big obstacle." Dr. Roy S. Herbst of the M.D. Anderson Cancer Center added that this could also pose a problem for researchers using adaptive trials to test predictive markers. This requires identifying the markers and then testing them in real time within the same trial, "so this whole idea of CLIA and how we're going to do it and get paid for it when the assays are being developed in real time is a pressing issue," he said.

TEST RELIABILITY

Even if all the obstacles above are overcome, and tests and clinical trials do reveal the analytical validity, clinical validity, and clinical utility of a predictive test, the reliability of test results can still be problematic due to

---

[2] *The Clinical Laboratory Improvement Amendments of 1988.* Public Law 100-578. (October 31, 1988).

inaccuracies in how the test is performed in the laboratory. Emblematic of these issues are tests for HER2 amplification.

The breast cancer drug Herceptin is only effective in women with tumors that have excess copies of the HER2 gene. When Herceptin was ready for clinical testing, a technique used to detect gene amplification called fluorescent in situ hybridization (FISH) was in its infancy and was not appropriate to use to detect HER2 amplification, said Dr. Mass. Instead, researchers at Genentech developed a test that used an immunohisto-chemical technology to detect HER2 protein levels, which, when elevated, indicate gene amplification (Mass, 2009). When Herceptin first came out on the market, its label specified that it be used in conjunction with this "HercepTest" diagnostic.

Shortly afterward, further tests by Genentech suggested that the FISH test for HER2 amplication was more accurate and reliable than the HercepTest. Four years later the FISH test entered the market, and was also added to the Herceptin label as an option for discerning patients likely to respond to the drug. However, for reimbursement and other reasons, the FISH test is often only done when the HercepTest test gives an equivocal result, so many more HercepTests than FISH tests are conducted, Dr. Mass noted.

Despite the break throughs in HER2 testing, lab testing errors can be as high as 20 percent even in CLIA-certified labs, according to a study done by the College of American Pathologists (CAP) and ASCO (Table 1) (Wolff et al., 2007). This suggests the need for better quality control and standardiza-

TABLE 1 HER2 Diagnostic Test's Error Rates (Concordance Central vs. Local Lab, Study N9831)

	JNCI 2002 (total n = 119)	ASCO 2004 (total n = 976)	JCO 2006 (total n = 2,535)
IHC 3+ (HerceptTest)	74%	79.5%	82%
FISH + (PathVysion)	67%	85%	88%

NOTE: ASCO = American Society of Clinical Oncology, JCO = *Journal of Clinical Oncology,* JNCI = *Journal of the National Cancer Institute.*
SOURCES: Shak (2009); Wolff et al. (2007).

tion in HER2 testing, Dr. Shak said. Ms. Bonoff added, "There's significant variation in the results of these commonly used HER2 tests in different laboratories, as well as different tests for the same marker, illustrating the crying need for standardization of testing parameters. As a patient advocate, I must point out how unnerving it is for patients when they face ambiguous and/or divergent results from predictive tests. We need the investment and policies that encourage bringing those technical innovations to standardized and practical implementation. The process of standardization is very important."

Dr. Hayes added that "the best marker stinks unless the assay is done well." He noted that the ASCO/CAP HER2 guidelines have led to CAP establishing proficiency requirements for HER2 testing. For a lab to achieve CAP accreditation for HER2 testing, it must achieve a 95 percent concordance with a central reading (CAP, 2007). He believes the FISH test's accuracy has been overestimated in comparison to HercepTest. "I think FISH has been done well because Mike Press does it well, and the people at Mayo Clinic do it well. But there are just as many mistakes in FISH as there are in HercepTest," he said.

TRANSLATION CHALLENGES

An additional technological hurdle to personalized medicine in oncology is implementing predictive tests into clinical practice. For example, an analysis by United Healthcare revealed that patients who are eligible for Herceptin often do not receive it, and those who are unlikely to respond to Herceptin are often treated with the drug (Phillips, 2008). This analysis estimates that as many as a third of patients may have received inappropriate treatment, Ms. Bonoff reported.

Ms. Bonoff was also critical of the shortcomings of Herceptin as a treatment for breast cancer. About a quarter of breast cancer patients overexpress the HER2 gene, and thus are eligible for treatment with Herceptin. Of those eligible, she said, about 5,000 U.S. patients receive Herceptin without any clinical benefit, and about 7,000 patients who could derive benefit are not being treated because of a false-negative test result (Phillips, 2008). Even patients who do respond to Herceptin eventually usually experience a recurrence of their breast cancer (Romond et al., 2005). "As patients, we have a tempered view of all the latest promises of breakthroughs of tests that will reduce our treatment, and rarely do; of new biomarkers that will make a real difference, and have not," she said. "Don't oversell personalized medicine. We know that breast cancer is many different diseases, and treatment that is tailored to specific tumor characteristics seems like a logical research path to follow. But we must remember that an intervention in the lab is years away from clinical impact. We are going in the right direction, but we should not jump the gun before the evidence is in. I am concerned about promising new approaches to diagnoses that are hyped before they are adequately validated or don't positively impact patients. The most elegant and innovative scientific research in the world means nothing if it can't help any person to live longer or better."

Ms. Bonoff also asked researchers not to neglect prevention in their efforts to develop personalized medicine. "Right now we have poor tools to determine who is at risk for developing disease, and end up applying a one-size-fits-all approach to most screening and prevention interventions. This results in overuse of medical resources and overdiagnoses," she said.

Contributing to the misuse of predictive tests is also insufficient physician education, Dr. Ratain pointed out. "The average clinician knows very little pharmacology and genetics, so how is he or she supposed to use pharmacogenetics?" Mark Gorman of the National Coalition for Cancer Survivorship asked, stating that ultimately the decision to use predictive tests will be made by clinicians and their patients. "There are policy ways to try and address the knowledge and skill of the clinicians, decision support, and the time that clinicians have to spend with their patients trying to support and sort through complicated bodies of information," he said. Ms. Bonoff also stressed the need to educate physicians about new developments in personalized medicine, questioning how quickly new treatment protocols are disseminated into the communities where most patients are treated. "Once we figure out which patients benefit from a specific treatment, when all the evidence is in, will we make the clinical changes necessary to make sure that only those patients receive treatment? How do we integrate new evidence into existing clinical practice?" she asked.

The Secretary's Advisory Committee on Genetic Testing (SACGT) (the predecessor to the current Secretary's Advisory Committee on Genetics, Health, and Society, or SACGHS) recognized that the clinical use of genetic testing could be improved by enhanced genetic education of healthcare providers, insurers, and patients (SACGT, 2000b). Clinical decision support tools, such as electronic medical records, might be able to fill in some of the gaps in that education, said SACGHS Chair Dr. Andrea Ferreira-Gonzalez of Virginia Commonwealth University. These tools can discern the information from a patient's record that will help physicians to make their clinical decisions. However, "it's not [known] how that would play out as we continue to leverage the information technology of the electronic medical record to start mining the data to not only improve [health care], but also improve the education of healthcare providers," Dr. Ferreira-Gonzalez said. SACGHS also recommended that the U.S. Department of Health and Human Services (HHS) allocate resources to the Centers for Disease Control and Prevention (CDC), Agency for Healthcare Research and Quality (AHRQ), Health Resources and Services Administration, and National Institutes of Health (NIH) for research and development of clinical decision support systems (SACGHS, 2008a). Dr. Ferreira-Gonzalez stressed that "you can do the testing, but if the clinician or the consumer doesn't know how to interpret the test, you might as well have not done the quality testing."

Dr. Ratain noted that clinicians will probably have to wrestle with data overload problems. The commercial software packages that clinicians typically use are not designed to reliably analyze and interpret the immense amount of data generated with genome-wide typing or sequencing. He also questioned the availability of these tests to clinicians at large. Dr. Johnson noted that there may also be limited availability of patient tumor tissue for such testing, especially for inaccessible tumors, such as lung cancers. Dr. Mass added that a biomarker study his company did on ovarian cancer required them to remove a large piece of tumor with a laparoscopic biopsy. It took a year to acquire the Institutional Review Board approvals for the protocol at the half-dozen sites in which they conducted the study.

CODEVELOPMENT CHALLENGES

Several speakers stressed the need to develop biomarkers concurrently with targeted drugs. Dr. Shak noted that it was not until Herceptin was in Phase III testing that a clinical assay was developed to identify people likely to be responsive to the drug, and "we scrambled over the last 9 to 12 months to find a commercial partner to work out what needed to be done in order to present data to the FDA regarding the HercepTest. An important lesson that I and many of us have learned is that you don't want to think about that late," but rather it is important to start developing a biomarker assay early on in the drug development process. Ms. Bonoff said, "Tamoxifen and Herceptin are perfect examples of how it's so important that the discovery of predictive biomarkers must not exist in a void, and that the successful development of drugs depends on the parallel development of predictive biomarkers. If we don't want drugs to be developed in a void, we must ensure that the interdisciplinary work needed becomes standard practice or we're just wasting time."

Dr. Hayes noted that the chances of codevelopment of a tumor marker and a therapeutic occurring at the start of clinical testing are about 10 percent, because often what was originally thought to be a good marker for the therapeutic turns out to be ineffective, and a new tumor marker shows more promise. He suggested that the FDA should stipulate that no registry trial be accepted without a prospective codevelopment plan, or at least a prospective plan for a specimen bank, and a transparent system to access specimens that provides adequate protection for intellectual property rights. "The sin is that the large pharmaceutical companies have not collected and bagged and stored specimens so that we could ask questions from the trials that they've run," he said.

"A lot of therapies are generic, like chemotherapy, that we apply right now based on prognostic factors," Dr. Hayes noted, "but we could really come up with better predictive factors for these therapies." He suggested that in addition to codevelopment of specific markers, testing of generic markers for existing chemotherapies should also be done. Dr. Leonard concurred, noting that "there is a tremendous amount of research on markers for the proper use of existing drugs. But if you're going to fix the marker development, validation, and implementation system for the new drugs, please do it for existing ones too." Dr. Bruce Quinn of Foley Hoag, LLP, added that biomarkers for generic drugs are just as important to develop as those for new branded drugs.

Regulation of Predictive Tests

The predictive tests used in personalized medicine are overseen by two federal agencies—the FDA and the Centers for Medicare & Medicaid Services (CMS). The Medical Device Amendments of 1976 to the Federal Food, Drug, and Cosmetic Act brought the marketing of devices, including in vitro diagnostics, under FDA regulation (hereinafter, in this report, "companion diagnostic tests.")[1] The FDA has exercised regulatory discretion with regard to laboratory-developed predictive tests (hereinafter, in this report, "laboratory-developed tests"), and does not oversee the development of these tests. The laboratories that provide these tests are, however, subject to oversight by CMS under CLIA, with the goal of ensuring quality laboratory testing services.

The FDA and CMS authority for the oversight of predictive tests are described in detail below. These sections are followed by a discussion on whether the current, dichotomous system is the best approach to overseeing these types of tests.

OVERVIEW OF THE FDA'S REGULATION OF PREDICTIVE TESTS

Dr. Alberto Gutierrez of the Office of In Vitro Diagnostic Devices (OIVD), FDA, explained how the FDA regulates companion diagnostics

---

[1] The Medical Device Amendments of 1976. Public Law 94-295. (May 28, 1976).

(predictive tests that have gone through the FDA approval process). He began his talk by pointing out that "in personalized medicine, the companion diagnostic really becomes key, because if you're going to be given a therapeutic, or you're going to be taking a clinical action based on the companion diagnostic, the diagnostic has to be right." The Medical Device Amendments of 1976 gave the FDA authority to regulate devices, including companion diagnostic tests, based on the amount of risk that is linked to the use of that device. Devices are classified into one of three risk categories (Classes I, II, and III), where Class I devices have the lowest level of risk and Class III devices have the highest.

The regulatory requirements necessary for approval of a device are based on the devices classification. Manufacturers of Class I devices, such as Band-Aids or pH tests, have to register their test with the FDA and follow general controls, such as adhering to good manufacturing practices, reporting device failures, and developing and using a system for remedying such failures (FDA, 2009b). The requirements for Class II devices are more complex. This is where most companion diagnostic tests fit into the classification scheme. Manufacturers of Class II devices need to follow FDA guidance documents that detail what manufacturers need to provide in order to receive FDA market clearance of their medical device, quality system regulations, and other special controls. They also must show that their device is substantially equivalent to a device that is on the market, or was on the market before 1976. This process is what the FDA calls "premarket notification (510(k))" (FDA, 2009b). Class III devices are the most complex and pose the highest degree of risk. Manufacturers of Class III devices are required to submit an application for Premarket Approval (PMA) to the FDA that details the safety and effectiveness of their device. The device cannot enter the market until after the FDA reviews and approves this application (FDA, 2009b).

In general, "the nice thing about this regulatory process is that it is quite malleable," Dr. Gutierrez explained. "We can apply the necessary regulation depending on both the risk of your test and its complexity, so it allows the reviewers the ability to mold their regulatory process to what you have." The FDA determines a device's risk classification based on the intended use of the device. If a device has more than one intended use, it will have a separate review process for each use. For example, "you could have a device that is used for monitoring cancer, which will have a lower risk than a device that does screening for cancer, because if you tell somebody they don't have cancer when, in fact, they do, you can actually put them at very high risk," Dr. Gutierrez said.

In its review of devices, the FDA considers analytic validity (the accuracy of a test in detecting the specific entity that it was designed to detect) and clinical validity (the accuracy of a test for a specific clinical purpose), but not clinical utility (the clinical and psychological benefits and risks of positive and negative results of a given technique or test). This means that although the FDA evaluates whether a companion diagnostic test can provide accurate information for clinical decision making, it does not thoroughly assess the risks and benefits of using the test on patients. However, Dr. Gutierrez added that sometimes the FDA consults with experts as to whether the risk of a device giving the wrong information outweighs the benefits of allowing the test.

In addition, the FDA regulates companion diagnostic tests by ensuring that all of the claims made on a diagnostic test's label are accurate and can be supported by evidence. The OIVD review of device performance is transparent with the reviews posted on the FDA website (FDA, 2009c). The FDA also does postmarket surveillance, and takes action to help resolve device failures when they are detected (FDA, 2009d).

In 2005 FDA published a white paper on codevelopment of diagnostics and therapeutics, and established a procedure whereby a codeveloped drug and companion diagnostic could undergo parallel FDA review and approval (FDA, 2005). This process has led to drugs receiving FDA approval based on studies that only tested the drug in marker-positive patients, rather than on unselected populations. However, there are shortcomings to this process, Dr. Gutierrez stressed. This type of testing does not conclusively show that the drug's effectiveness is linked to the companion diagnostic test result because this conclusion can only be determined by testing the drug in both marker-positive and marker-negative patients. "What we learned from HER2 is that if you do a trial in which you actually have only marker-positive patients, in the end you actually know the positive predictive value of the test, but not much else about the test," he said. Such a study does not indicate sensitivity, specificity, and the negative predictive value. This poses problems when a competing biomarker is discovered because its comparative value to the older test cannot be fully ascertained given the lack of information on the marker-negative population. However, one participant noted that it would be difficult, if not unethical, to accrue patients who are marker negative to a clinical trial of a targeted agent because they are not likely to receive any benefit.

OVERVIEW OF CMS'S REGULATION OF LABORATORIES PERFORMING PREDICTIVE TESTS

As discussed above, laboratory-developed tests are not currently regulated by the FDA, however the agency has the authority to do so. Instead CMS regulates the laboratories that develop these tests through CLIA. Laboratory-developed tests have historically been conducted in a single laboratory on specimens that come from the nearby patient population. More recently, laboratory-developed tests are being done in a single lab using samples from all over the country. In contrast, FDA-approved companion diagnostic tests are generally developed by industry to be used in laboratories throughout the country, if not the world. FDA companion diagnostic "test kits" need to be more robust because anyone can buy the kit and perform the test, regardless of their expertise, Dr. Leonard said.

The FDA has chosen not to regulate laboratory-developed tests because of a lack of resources, according to Dr. Steven Gutman of the University of Central Florida and former head of OIVD, and not necessarily because they pose less risk than FDA companion diagnostics. When the decision was made not to oversee laboratory-developed tests, most of these tests were for research purposes and were not commercialized. However, today many companies are using laboratory-developed tests because they present an easier method of getting predictive tests on the market.

Dr. Penelope Meyers of CMS explained that there are no CLIA-certified or approved tests because CLIA certifies and regulates the laboratories doing the testing, and not the tests themselves. A laboratory must receive CLIA certification to be reimbursed by Medicare/Medicaid. In addition to CLIA requirements, some laboratories must follow more stringent rules required by certain states for licensure and permits. The purpose of CLIA was to ensure that patients receive the same quality of laboratory testing, regardless of where the test is performed, be it in a hospital laboratory, a large reference laboratory, or a physician's office, said Dr. Meyers.

CLIAS's complexity requirements apply to all clinical laboratories and their stringency depends on the complexity of the testing being performed (high, moderate, or waived). FDA risk assessment is based on the impact of the test result on decision making (low, moderate, and high risk based on impact on patient.) Laboratories that perform testing with the lowest degree of complexity (waived testing) are not subject to any routine CLIA oversight, and do not get inspected. These laboratories are only required to follow the manufacturer's test instructions. Genetic-based predictive tests are all considered to be high complexity. Laboratories that perform these tests must follow the most stringent CLIA requirements, including the most rigorous personnel requirements. However, there are no specific personnel requirements for genetic-based predictive tests.

CLIA oversight focuses on laboratory procedure, on the training of laboratory personnel, and on the credentials needed for test interpretation. CLIA oversees the registration, certification, and accreditation of laboratories, proficiency testing of the lab, regulations governing the physical plan, record retention, and other facility requirements and quality control systems.[2] CLIA regulations require 84 listed analytes to have proficiency testing; however, none of the listed analytes is for a genetic test. A current CDC project is attempting to update the analyte list, partly in response to the issues raised by the genetic testing community about proficiency testing, Dr. Meyers said.

CLIA preanalytic requirements include those governing specimen submission and handling. The major postanalytic requirement is that laboratories have systems for collecting, responding to, and acting on communications and complaints about performed tests. All laboratories, except for those doing waived testing, are inspected every 2 years, and CLIA can take enforcement action against labs that do not correct deficiencies detected during inspections.[3] It is important to note that CLIA does not have any legal mechanism to enforce CLIA violations in laboratories performing predictive tests that do not have CLIA certificates. However, there have been instances where CMS CLIA offices have sent their state and regional office surveyors into unregulated laboratories that were known to be conducting predictive tests, and have been successful in having them either apply for a CLIA certificate and submit to inspection or cease testing, Dr. Meyers said.

CLIA has a lengthy list of requirements that ensures the analytic validity of the testing a laboratory performs, but CLIA does not regulate the clinical validity of a test, unlike the FDA's regulation of companion diagnostic tests. Clinical research data are not required to support the claims on the laboratory-developed tests' label, even if these tests are linked to the same degree of risk and complexity as companion diagnostic tests approved by the FDA. In addition, there is no requirement for reporting adverse events with laboratory-developed tests, nor is there

---

[2] 42 C.F.R. Ch. IV Part 493. http://ecfr.gpoaccess.gov/cgi/t/text/text-idx?c=ecfr&sid= 24b0d768489ac16c021cd4c5656568b4&rgn=div5&view=text&node=42:4.0.1.5.29&idno=42.

[3] 42 C.F.R. Ch. IV Part 493, Subparts Q and R.

public information on their analytic or clinical validity, as there would be for tests that undergo FDA reviews.

"There is regulation that requires a laboratory director to offer testing that is appropriate for the patient population, and this rule is sometimes interpreted loosely to mean that the clinical validity of a test needs to be considered," said Dr. Meyers. "But CLIA really does not directly regulate clinical validity, and we don't require specific data on clinical validity for laboratory-developed tests." If CLIA surveyors notice anything questionable about a laboratory-developed test during an inspection, they can consult with an expert at CMS, CDC, and FDA, Dr. Meyers said.

When a laboratory is going to implement an FDA-approved or -cleared companion diagnostic test, the laboratory must verify that the test's performance specifications are met. But for laboratory-developed tests, the laboratory establishes its own performance specifications, such as analytical sensitivity, specificity, and other performance characteristics required for test performance, and can begin offering the test once a laboratory director deems these performance specifications suitable.

SHOULD THE FDA DO MORE?

Some companies have tried to use the CLIA regulatory pathway for their predictive tests inappropriately, rather than go through the FDA approval process. For example, LabCorp put its OvaSure test on the market as a laboratory-developed test, even though it was actually an in vitro diagnostic test according to the FDA. Eventually, LabCorp pulled OvaSure from the market due to FDA pressure. Recognizing this confusion in the regulation of laboratory-developed tests, the FDA published the Analyte Specific Reagent rule in 1997 and again in 2007, which specified that the materials used in laboratory-developed tests must follow FDA rules for Class I devices (FDA, 1997, 2007a). More recently, the FDA began working on the In Vitro Diagnostic Multivariate Index Assay (IVDMIA) guidance for high-risk, high complexity tests, such as Oncotype Dx (FDA, 2007b). This guidance defines and specifies the regulatory status of IVDMIAs, and clarifies that even when offered as laboratory-developed tests, IVDMIAs must meet pre- and postmarket device requirements, including premarket review requirements in the case of most Class II and III devices. However, the FDA is not currently enforcing the IVDMIA guidance and it is unclear when or if it will be finalized.

In addition, several speakers and discussants noted that these guidance documents are insufficient to counter the lack of regulatory parity between FDA-reviewed companion diagnostic test kits and those laboratory-developed genetic tests that fall under CLIA's purview. "Industry sees a big disparity between those tests that go to market as a laboratory-developed test, and what they have to do to get an FDA-approved or -cleared test," said Dr. Gutierrez. Genentech recently filed a citizen's petition with the FDA that asked the agency to review its regulation of laboratory-developed tests, with the aim of having all tests that are used or intended to be used for therapeutic decision making undergo the same scientific and regulatory standards (Genentech, 2008). "There's been a broad proliferation of assays that are allegedly being used to make decisions about patient care without any type of FDA clearance as it relates to efficacy or safety," said Dr. Mass. "We think that a lot of these claims are misleading, or certainly unsubstantiated by the kind of data that would be required of a drug manufacturer to get marketing approval for a drug."

An example of such unsubstantiated claims is a predictive test used to determine the likely responsiveness of lymphoma patients to rituximab therapy. The maker of this test claims it will enable physicians to "confidently predict" whether lymphoma patients will respond to rituximab (PGxHealth, 2009). This claim was supported with data generated by the company that devised the test, which showed that when individuals with follicular lymphoma are homozygous for a specific gene, they will have a 100 percent response rate to the drug, whereas those who are heterozygous or completely lack the gene will only have a 67 percent response rate (Figure 6) (Cartron et al., 2002; PGxHealth, 2009). "The confidence intervals here are quite wide and overlapping so one could really question whether the claims being made by this assay system are relevant," said Dr. Mass.

Genentech did its own analysis of the test on patients with diffuse lymphoma, and found that when rituximab was added to the cyclophosphamide, doxorubicin, vincristine, and prednisone (CHOP) chemotherapy protocol sequentially in a few hundred patients, the tumors of 40 percent of those homozygous for the gene progressed on the regimen, compared with 57 percent of those that were heterozygous or completely lacking the gene—a much less striking difference, Dr. Mass noted (Vose et al., 2009). A larger study may have revealed more differences between the homozygous and heterozygous patients. However, Dr. Mass said, unlike the univariate analysis done by the maker of the test, Genentech did a more statistically rigorous multivariate analysis in which it corrected for other prognostic variables not considered in the simpler analysis. "We don't think physicians or patients

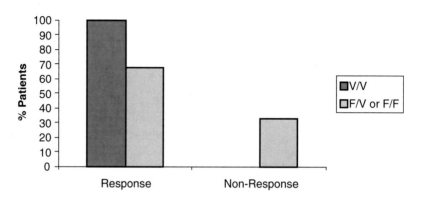

FIGURE 6 Data from PGxHealth on follicular lymphoma patients receiving Rituximab monotherapy. This study shows that when individuals with follicular lymphoma are homozygous for a specific gene, they have a 100 percent response rate to Rituximab, whereas those who are heterozygous or completely lack the gene only have a 67 percent response rate. Response is defined as complete or partial response, and non-response is defined as stable or progressive disease by Cheson criteria.

NOTE: F/F = lack the gene, F/V = heterozygous for the gene, V/V = homozygous for the gene, n = 10 for the V/V group, n = 39 for the F/V or F/F group.

SOURCES: Mass (2009); Cartron et al. (2002); PGxHealth (2009). This research was originally published in *Blood.* Cartron, G., L. Dacheux, G. Salles, P. Solal-Celigny, P. Bardos, P. Colombat, and H. Watier. Therapeutic activity of humanized anti-CD20 monoclonal antibody and polymorphism in IgG Fc receptor Fcgamma RIIIa gene. 2002; Vol 99(3):754–758. © the American Society of Hematology.

should be subjected to this test without more rigor around the claims being made about it," he said. "We think that any test that's making a claim about clinical effectiveness should be reviewed by the FDA."

Dr. Mass also questioned the claim that laboratory-developed tests do not have to be reviewed by the FDA because they tend to not pose safety hazards. In the context of predictive tests, safety can be defined as the right patient getting the right drug, and the wrong patient not getting the wrong drug, he noted. However, there are limited examples of these types of safety issues being considered for laboratory-developed tests because CLIA does not require this type of record keeping. As a result, it may take years for safety problems in laboratory-developed tests to become apparent, he pointed out. "Did Mrs. Jones actually get the right therapy based on some assay that was conducted, and was her outcome altered in some way by that treatment?" To ensure tests are safe, there needs to be some clinical validity or utility measurements, he said.

Dr. Gutierrez concurred that predictive tests that are intimately tied to a therapeutic should be approved by the FDA, whether or not they are laboratory developed, because "for the drug to be safe and effective, the device itself has to be controlled." Yet several tests that are intimately tied to therapeutics did not undergo FDA review, including Oncotype DX.

FDA regulation of laboratory-developed tests might stifle innovation and prevent the iterative development of these tests that often occurs, Dr. Gutierrez pointed out, but Dr. Mass disagreed that this would be problematic. Dr. Mass recognized the lack of resources that currently prevents FDA from reviewing laboratory-developed tests, but added, "If we believe as a community that this needs to happen, there are certainly ways that the resourcing can be applied to review the tests that we think are important to review. The CLIA process is essential in terms of laboratory quality, but it's not really closing the loop on proper clinical validation. If we can't improve this regulation, we can never fully realize the promise that personalized medicine should bring to patients."

An additional reason for providing more comprehensive regulation of predictive tests is that drug developers want more predictability, Dr. Gutierrez said. Ms. Stack called for increasing clarity on both the regulation and reimbursement of predictive tests so venture capitalists, such as herself, can continue to develop innovative companion diagnostic companies. "I don't want more regulation. I just want clarity because it's really hard to develop a business when you're not clear of how it's going to be regulated and reimbursed," she said.

IS THE STATUS QUO APPROPRIATE?

Some speakers and participants questioned whether FDA review of laboratory-developed tests would be sufficient. "Everybody talks about the FDA pathway as the gold standard ideal, but there are lots of problems with the FDA process," said Dr. Leonard. "Nor do I believe that the CLIA process is perfect," she added. Dr. Hayes said that when he and others develop ASCO guidelines, "We don't care if the FDA has or has not approved a test, because FDA approval doesn't mean the test should be used to take care of patients, and possibly the best test we have in breast cancer now [Oncotype DX] has never been approved by the FDA. The whole system needs to be revamped. We need to review tests the same way we review drugs." Ms. Bonoff concurred, saying, "The controlled, randomized clinical trial has been the gold standard for drug treatment. Why shouldn't that also be the standard for tests that will guide the use of therapy? We really need to know that the tests are dependable because we're making decisions that affect our lives based on these tests. The current system makes no sense. Predictive tests are now in the clinic without FDA review, and FDA-reviewed companion diagnostic tests may be used for non-approved means. There is an essential need for stronger criteria and oversight to replace the current patchwork system."

Dr. Leonard agreed that "the systems need to be tweaked to ensure greater safety and demonstration of efficacy where there are holes." However, she expressed concern that making every laboratory-developed test go through an FDA review process would slow down the development and use of these tests. The pathway for a laboratory-developed test to enter the market is rapid, whereas the FDA-approved test pathway is much slower. "Please don't eliminate the rapid pathway for test availability. That would be throwing the baby out with the bathwater," she said. Instead, she suggested that FDA review should be done on those tests with high complexity, and for those tests being performed by institutions on patient specimens collected from outside the local community. An institution does not have the same degree of control over commercialized assays when specimens are coming from all over the country or world.

Ms. Gail Javitt of the Genetics and Public Policy Center, Johns Hopkins University, said, "It's wrong to think that a one-size-fits-all approach to the regulation of all laboratory-developed tests and genetic tests would work." She suggested having different regulatory pathways based on risk. In 2008, the SACGHS recommended that HHS convene a multistakeholder, public- and private-sector group to develop criteria for determining the appropriate oversight of laboratory tests, and a process for systematically applying the criteria, said Dr. Ferreira-Gonzalez (SACGHS, 2008a). "Before increasing oversight, the benefits and harms to patient access and cost should be considered," she said.

However, SACGHS's predecessor, SACGT, recommended premarket review for all predictive tests, including laboratory-developed tests, said Dr. Wylie Burke of the University of Washington, and former member of SACGT (SACGT, 2000a). It suggested that premarket review should be streamlined, and a template should be developed to standardize the FDA review process for discerning tests that are more complex. The primary goal of this template-driven approach is accurate labeling, according to

Dr. Burke. Labeling of predictive tests should specify the intended use of the test, the specific actions that will follow from use of the test, the clinical condition for which the test is performed, as well as specificity, analytic validity, and any known clinical validity and clinical utility. Transparency about the current state of knowledge for a test might provide protection against unsafe testing, Dr. Burke said.

When it comes to ensuring the quality of the testing itself (analytic validity), Dr. Leonard stressed that the quality of testing done by CLIA-certified labs through laboratory-developed tests is virtually the same as that for FDA-reviewed tests. FDA reviews of tests are done using the same template for providing standard information on a companion diagnostic test as used by laboratories under CLIA, and the proficiency testing is basically the same for both laboratory-developed tests and FDA-regulated tests. One analysis found a 98.1 percent accuracy with samples sent to laboratories for proficiency testing for a broad range of genetic tests, most of which are laboratory developed (SACGHS, 2008b). This is comparable to the 97.6 percent accuracy in proficiency testing done for HIV-1 screening and cardiac markers for heart attacks, Dr. Leonard noted.[4]

Dr. Leonard added that there is no evidence that laboratory-developed genetic tests are of poorer quality than other laboratory tests. The SACGHS report concluded that genetic testing is not an exceptional type of laboratory test (SACGHS, 2008b). "For the purposes of oversight of genetic testing, that is, analytical validity and clinical validity, we consider that genetic and genomic testing is not different from other testing we do in the laboratory," said Dr. Ferreira-Gonzalez. "There are quality issues across the board," Dr. Leonard added. "Stop focusing on laboratory-developed tests and genomic tests as special. There are problems with tests that we currently do that aren't laboratory-developed tests or aren't genetic tests. We need to focus on the quality and proper use and interpretation of *all* tests," she said.

Dr. Leonard and other participants at the workshop claimed that the main quality issue that needs to be addressed better in regulation is the need to show clinical utility, which neither the FDA nor CLIA requires. "We need to know [whether] these tests [are] usefully affecting the outcome of patients in the clinic. Neither of these processes gets at that," Dr. Leonard said. However, Dr. Mass pointed out that determining clinical utility is difficult to do because "there's not consistency of what clinical benefit really means as yet." Dr. Ralph Coates of the CDC added that a 2007 IOM report

---

[4] CAP Proficiency Test Results Summary for these tests from 2008 data.

on biomarkers called for defining the translation pathway for biomarkers more clearly (i.e., determining what information on clinical validity and utility is needed, and what kind of research should be done to acquire this information) (IOM, 2007). He noted that in research, "there seems to be more interest in novel findings—what's new—rather than summarizing what we really do know and don't know." He suggested there should be more systematic evidence reviews, and support for research addressing knowledge gaps identified from those systematic evidence reviews.

In addition, personalized medicine and predictive tests are rapidly evolving, and need to be continuously evaluated for testing outcomes, said Dr. Ferreira-Gonzalez. Dr. Herbst added that "we're dealing with new information that changes maybe not the analytic validity of the test, but the clinical validity and ultimately probably the clinical utility." Dr. Gutierrez pointed out that the way the FDA deals with new information is to specify what is currently known and unknown on the label of a diagnostic or therapeutic. "The iterative nature of this is very challenging, and we have a mechanism for getting products out in the market quickly, but if it's investigational, we think patients deserve to know," he said. "When we know something's a winner—it's been hit out of the ballgame—we try to clear or approve it. When we know it's a loser, we lie down like we are in front of the railroad tracks trying to block it. And when it is in this gray zone, we try to label it." An example of this regulatory behavior was pointed out by Dr. Amado, who noted that the FDA has agreed to include information about the lack of activity of anti-EGFR antibodies in the setting of KRAS mutations in the labels (i.e., while the indication remains broad, the label states that in a retrospective analysis, patients with KRAS-mutant colorectal tumors did not benefit from panitumumab or cetuximab, drugs that target the EGFR) (Amgen, 2008; ImClone Systems, 2008). This explicit labeling was, in part, a compromise between the test developers, who thought it was not feasible to do a prospective analysis, and the FDA, who wanted prospective data to evaluate the test.

POLICY SUGGESTIONS

In addition to suggesting that the FDA review all laboratory-developed tests or all complex tests, speakers and discussants made several suggestions for improving the regulation of predictive tests. These suggestions included

- strengthening the proficiency requirements for laboratory personnel;
- increasing the transparency of data collected on laboratory-developed tests;
- restructuring and coordinating the oversight of companion diagnostic tests;
- improving FDA and CLIA enforcement of predictive test regulations; and
- assessing the clinical utility of predictive tests before or after they enter the market.

Improve Laboratory Proficiency

Laboratories performing predictive tests must enroll laboratory personnel in proficiency tests specific to the subspecialty of the tests they will be evaluating. CLIA requires proficiency testing of personnel at least once every 2 years for non-waived tests. However, a major deficiency of CLIA is that it does not require proficiency testing for all tests. Dr. Leonard suggested that one way to improve the regulation of laboratory-developed tests is to require stricter proficiency qualifications and personnel qualifications for predictive tests. She also suggested requiring proficiency testing for any test performed in the laboratory, regardless of whether the tests are FDA approved. Dr. Hayes pointed out before CAP proficiency testing was implemented for HER2 tests, 15 to 20 percent of HER2 analyses were done incorrectly in CLIA-certified labs. "As we began to have CAP proficiency testing, where your feet are put to the fire every 6 months, we've seen agreement go from 65 to 70 percent to close to 90 percent," he said.[5]

Ms. Javitt described a longstanding concern about the lack of mandatory proficiency testing for genetic tests because there is no specialty for them under CLIA. Dr. Ferreira-Gonzalez added that legally, laboratories are required to perform proficiency testing on only 84 analytes. SACGHS debated whether or not to recommend regulation to require proficiency testing for all analytes for which proficiency testing material is available. However, this requirement would have been problematic because the analytes needed to do proficiency testing for genetic tests are often unavailable, and genetic tests are a rapidly moving target. Consequently, SACGHS decided to recommend that HHS fund studies to assess alternative ways to

_____

[5] 42 C.F.R. Ch. IV Part 493, Subparts G and H.

conduct proficiency testing for genetic tests, including splitting samples with other laboratories or retesting one's own samples. SACGHS also recommended that HHS ensure funding for the development of certified or validated reference materials that can be used to validate assays. In addition, it recommended increased funding for the development of assay, analyte, and platform validations that could be used for quality control assessment and standardization of testing among different laboratories (SACGHS, 2008b).

The FDA Center for Devices and Radiological Health (CDRH) has developed many guidance documents on the acceptable levels of performance characteristics for predictive tests. Dr. Ratain suggested that insurers could require that predictive tests be performed in labs that meet the standards specified in these guidance documents, regardless of whether the tests are subject to FDA approval.

Increase Transparency

Ms. Javitt stressed the importance of making proficiency data public. Currently, the CAP collects and compiles these data, but does not release the data to the public. "There should be a way for the public to access proficiency testing data so that they can make decisions about laboratory quality," she said. Dr. Leonard agreed there should be transparency in how laboratories operate and perform, and added that she understood that under CLIA, CMS is directed to make proficiency testing results public, but it does not currently comply with this requirement.[6]

The need for transparency of data about predictive tests, including data that have been traditionally considered industrial "trade secrets" not to be divulged to the public, was stressed by Robert Erwin of the Marti Nelson Cancer Foundation. "We are missing huge chunks of information, and the quality of decisions depends a lot on the quality of information going into making those decisions," he said.

The SACGT report recognized the need for more transparency in data collected on predictive tests, and recommended that genetic test developers be required to provide information on analytic validity, clinical validity, and clinical utility, said Dr. Burke. The committee knew that data on

---

[6] Section 353 of the Public Health Service Act, Section f on Standards, #3 on Proficiency Testing, part F on page 228, states that Proficiency Testing results must be publicly available.

clinical validity and clinical utility was likely to be very limited (SACGT, 2000a). However, "the point was that there should be transparency. Manufacturers should provide what they know and what they don't know, including citations to the literature," she said. SACGHS subsequently also recommended that HHS appoint and fund a lead agency to develop and maintain a mandatory, publicly available, Web-based registry for laboratory tests. It directed that a committee of stakeholders should determine what information should be entered into this registry (SACGHS, 2008b). The Twenty-First Personalized Medicine Coalition and the American Clinical Laboratory Association have also both supported the idea of registering laboratory-developed tests.

A mandatory registry of laboratory-developed tests offers a number of benefits. It would help foster truth in labeling, Dr. Ferreira-Gonzalez said. In addition, Dr. David Parkinson of Nodality, Inc., noted that this sort of registry would be extremely helpful in discerning the underlying biology governing the effectiveness of biomarkers and targeted drugs. "So much of what I hear about in the biomarker world is isolated tests, one point in time. The real information and meaning of these tests come when you are actually following patients longitudinally and when there's some sort of intelligent life force looking at the result of the biological characterization, the therapeutic action, and the outcome. Then you start to understand what the biology means," he said. Dr. Herbst added patients and physicians both feel a great deal of confusion or lack of knowledge about recently developed predictive tests. He promoted the development of a registry that would provide real-time information so that patients could be treated in the best possible way.

Restructure and Coordinate Oversight

To address the problem of having two independent regulatory paths for predictive tests under the FDA and CLIA, several presenters and participants suggested ways to restructure and coordinate this oversight. Dr. Darryl Pritchard of the Biotechnology Industry Organization (BIO) suggested reorganizing CLIA as a part of the FDA, and therefore under the same leadership structure to enable a system-wide approach to addressing regulatory gaps. However, Dr. Gutman replied that "it would be both an administrative and statutory challenge to do so because CLIA and the FDA are administratively and legally driven by widely different starting points. Although if you want to think out of the box and push this, it certainly strikes me that anything is possible if you're trying to fix a broken system."

Dr. Hayes suggested that all oncology regulatory activities within the FDA—both devices and therapies—be consolidated under a single branch or committee. Dr. Mansfield pointed out that the CDRH and the Center for Drug Evaluation and Research do have an intercenter oncology working group that considers both cancer diagnostic and therapeutic issues (FDA, 2009a). However, Dr. Ferreira-Gonzalez noted that the review by SACGHS uncovered a number of duplicate efforts assessing how to improve the oversight of genetic/genomic technologies among government agencies and offices within HHS. "They were not talking to each other. In some instances, they were doing exactly the same fact finding without even sharing some of the information," Dr. Ferreira-Gonzalez said. This discovery of bureaucratic redundancy led SACGHS to recommend that the Secretary of HHS should coordinate efforts in personalized medicine within the agency, and consider creating a new HHS office for this purpose (SACGHS, 2008a).

"There are several solutions," Dr. Gutierrez stressed. "But if you don't have the FDA doing the regulation, you're going to have to come up with a way to do it that makes sense, that people actually believe in, that is independent of both the laboratories and the manufacturers, and that is credible." Peter Collins of DxS Ltd. pointed out that globally, FDA regulation is seen as the gold standard to emulate, and any changes to that regulation would have global implications.

Improve Enforcement

Dr. Ferreira-Gonzalez suggested that the current regulations for predictive tests should be more uniformly enforced. She noted that when SACGHS solicited comments from stakeholders, a number suggested the need for better enforcement of current regulations related to laboratory testing. "Sometimes some of the problems we see are due to the regulation not being fully enforced," she said. Consequently, SACGHS recommended that the gaps in the enforcement of existing regulations for analytic and clinical validity be identified. For example, CLIA surveyors cannot inspect and close down laboratories that are not CLIA certified. They are restricted to providing information about a laboratory to the Government Accountability Office, and must rely on this office to take corrective action. SACGHS also recommended that CMS be empowered to take direct enforcement actions against labs that perform clinical tests without proper CLIA certification, including those that offer direct-to-consumer testing (SACGHS, 2008b). In addition, increasing the enforcement discretion of the FDA for laboratory-developed tests would not necessarily require new legal authority, Dr. Mansfield noted.

Assess Clinical Utility

Many participants and speakers acknowledged the need to assess the clinical utility of predictive tests. "Clinical utility today is being used by third-party payers to reimburse the testing that we do, but what we're starting to realize is that we don't have a lot of clinical utility data because we don't have the infrastructure to see what information is needed, and fund the collection of that data," said Dr. Ferreira-Gonzalez. Neither laboratories nor manufacturers have the resources to assess the clinical utility of genetic tests, for example, by building on the CDC's Evaluation of Genomic Applications in Practice and Prevention (EGAPP) initiative (SACGHS, 2008b). The Medicare Evidence Development and Coverage Advisory Committee (MEDCAC) identified clinical utility as a key issue for evaluation in Medicare coverage decisions, and recommended that Medicare use EGAPP methods in its evaluations of genomic tests (CMS, 2009c). SACGHS recommended that HHS create and fund a public/private entity to assess the clinical utility of predictive tests and develop a research agenda to address gaps in knowledge. SACGHS also recommended that HHS conduct public health surveillance to assess health outcomes or surrogate outcomes, practice measures, and the public health impact of predictive testing (SACGHS, 2008b). In addition, SACGHS recommended that researchers develop more evidence of clinical utility in genetic tests (SACGHS, 2006, 2008b).

Dr. Coates noted that research to assess the population health benefit of genetics or genomics-based tests or treatments comprise less than 3 percent of all published genetics research (Khoury et al., 2007; Woolf, 2008) (Figure 7). "The CDC is currently collaborating with the NCI to assess cancer genomics funding in specific, and how much of it is used for discovery, versus application, versus assessing the population health benefit of new genomics applications." A separate CDC/NCI effort on the comparative effectiveness of cancer care and prevention included the statement: "To date, there's been no systematic research conducted to compare the clinical effectiveness and cost effectiveness of cancer care and prevention based on genomic tools and markers compared to existing standards of care and prevention" (NCI, 2009a).

Stakeholders at an NIH/CDC Personal Genomics Workshop in

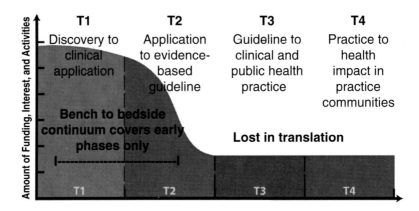

FIGURE 7 The research community's interest in implementation processes wanes along the continuum of cancer translation research. Ninety-seven percent of genetics research is published in the T0 and T1 phase.
SOURCES: Coates presentation (June 8, 2009); Khoury et al. (2007); Woolf (2008). Adapted from Khoury, M. J., M. Gwinn, P. W. Yoon, N. Dowling, C. A. Moore, and L. Bradley. 2007. The continuum of translation research in genomic medicine: how can we accelerate the appropriate integration of human genome discoveries into health care and disease prevention? *Genetics in Medicine* 9(10):665–674. Reprinted with permission of Wolters Kluwer Health.

December 2008 agreed that more multidisciplinary research should be done to fill knowledge gaps on the clinical validity and utility of predictive tests, Dr. Coates noted. (Khoury et al., 2009b; NCI, 2008). Participants also recommended that both personal and clinical utility be assessed, and that researchers should link science to evidence-based recommendations (NCI, 2008). The CDC, with the NIH and other organizations, has initiated the Genomic Applications in Practice and Prevention Network (GAPPNet) to increase communications among stakeholders to improve translation of genomic applications (Khoury et al., 2009a). Relevant stakeholders include test developers and others doing translational research; those developing evidence-based recommendations by linking evidence to practice guidance in a transparent and credible way; practitioners in clinics, public health, and community practice; and patient advocates.

Two examples from EGAPP that highlight the need to do more translational research evaluating the clinical utility of predictive tests are the recent evaluations of (1) breast cancer gene expression profiles, and

(2) the UGT1A1 genotyping. EGAPP used the systematic, evidence-based process it developed for evaluating predictive tests, and other applications of genomic technology in transition from research to practice, to evaluate these two applications of genetic/genomic technology (Teutsch et al., 2009). EGAPP found insufficient evidence to make a recommendation for or against the use of tumor gene expression profiles to improve outcomes in women with Stage I or II node-negative breast cancer (AHRQ, 2008; EGAPP Working Group, 2009a). EGAPP's evidence review found adequate evidence of the clinical validity of Oncotype DX and MammaPrint, but inadequate evidence of the clinical utility of Oncotype DX and no evidence of the clinical utility of MammaPrint. The analytical validity for both tests was also inadequate. Similarly, EGAPP did not find sufficient evidence to make a recommendation for or against the routine use of UGT1A1 genotyping in metastatic colorectal cancer patients, said Dr. Coates (EGAPP Working Group, 2009b; Palomaki et al., 2009).

However, one participant noted that guidelines and research reviews, such as EGAPP, are often based on a higher degree of efficacy than what doctors use for their clinical decisions. Dr. Hayes suggested that for a test to be clinically useful, its effect must be large enough that clinical decisions based on the test results have acceptable outcomes. These outcomes include cure, improved survival or palliation, or decreased exposure to toxicity due to useless therapy. The most useful tumor markers indicate those patients whose prognosis is so good or so bad that they are not likely to experience improvement with treatment, he said. In these patients the risks of therapy outweigh the benefits.

Ways to Capture Clinical Utility Data

Several suggestions were offered on how to improve the collection of clinical utility data needed to fully evaluate tests. Dr. Leonard suggested creating a registry of test and treatment outcomes, akin to the evidence development that CMS requires for treatments it considers investigational. During the data collection phase, the treatment or test would be reimbursed by Medicare/Medicaid. "This would allow new tests to get out there in medical practice only if you're collecting data on them," she said. Such collection and analysis of data would be aided greatly by a national electronic health record system, which would allow real-world data to be collected and used to determine clinical utility. "I would argue that you have to start at the bedside before you go to the bench, if you're doing healthcare research, because you have to have research that's driven by clinically relevant questions," Dr. Leonard said. If the data do not indicate clinical utility, the predictive test should be taken off the market.

Conceptually, such a registry would be more comprehensive than the registry of predictive tests proposed by SACGHS or SACGT. Dr. Leonard said she envisions a registry that would be disease based, and include all patients receiving the test or treatment, including all age, racial, and ethnic groups. "One of the problems when we do clinical trials is that often they don't mimic what we do in real clinical practice," she said. "The registry would create a pathway to acquiring clinical validity and utility data without the need for randomized, controlled clinical trials that are very costly, and, everyone agrees, would be very difficult to do for diagnostics."

Dr. Hayes pointed out that the proposed registry data would be confounded by the information doctors are given about the tests. Some patients would do well with or without treatment, and a test that predicts they would do well with treatment might lead physicians to assume they subsequently did well because of the treatment. But Dr. Leonard argued that physicians are now using tests for clinical decisions despite the lack of data on how clinically useful the tests are, and the advantage of the registry is that these data would be collected and analyzed. "In our current way of doing it, we don't get any of that data back," she stressed. "It's just not a good data collection process, and we may be able to speed it up if we did have some data collection." Data analysis for the registry may have to be done differently than for a randomized, controlled clinical trial, she added, because all of the potential confounding factors could not be controlled.

Dr. Austin countered that "health professionals have lots of hypotheses out there, and some prove to be right, and some prove to be wrong. But we don't know which ones are right and which ones are wrong until we do tests with proper control groups. So why not just do these studies? Why put things out on the market and then try to finagle proper control groups around it, which is very hard to do. What scares me is when doctors come into my lab and say 'I want this test' because of some paper they read that was probably of a study performed on just 23 subjects."

Dr. Leonard disagreed, pointing out the repetitive nature of medical practice. For example, she noted that if the main mutation in cystic fibrosis (CF) had not been discovered and a test for it put into practice, researchers would never have uncovered additional mutations that can cause CF, nor would those with the first mutation have been helped. "Some knowledge does allow us to move forward," she said. "There are processes that are alive and well and functioning in medicine, and I don't know that they're necessarily bad."

The FDA could play a serious role in safety determinations of predictive tests, Dr. Friend observed, while still allowing a dynamic, iterative process for determining clinical utility. "Imagine a world where not thousands of patients were enrolled in trials, but millions, and they involved drugs that were already approved," he said. "Even though a drug was first given for only one indication, 2 years from when it was first approved, the evidence-based data could come back in and change the indication," without having to undergo the lengthy process that is currently involved in having a drug label changed. For this to occur, patient advocates "would have to step up and say we need care that's more personal," and encourage more patients to enroll in trials, he said. Dr. Mass added that the potential evidence that could be developed by a registry similar to what Dr. Leonard proposed is great, but he stressed that the FDA should be involved to ensure postmarketing data are collected. "You'd need to have some leverage, and I don't think CLIA could do that," he said.

Another issue is off-label use of drugs or tests. Dr. Hayes noted that such use makes it difficult to accrue patients to trials assessing new indications. For example, it took 13 years to accrue enough subjects to conduct a randomized, controlled trial on the use of the prostate-specific antigen (PSA) test for prostate cancer screening because the PSA test was already approved for monitoring the progress of prostate cancer (Andriole et al., 2009; Schroder et al., 2009). "Because the assay was out there and being used, it took much longer to get the data we wanted," he said. The off-label use of a questionably useful test also made the predictive test regulation appear inconsistent, Dr. Quinn said.

Off-label use of tests can also be risky for patients, stressed Mr. Collins of DxS Ltd. The use of a test without sufficient evidence can lead to patients being denied treatment based on an unproven test that indicated the individual was unlikely to respond to the drug. "There has to be a better mechanism for dealing with this," he said. Dr. Leonard pointed out that European countries with universal health care coverage control test use based on evidence. "It's not that you don't get the test paid for. You don't get the test period" if there's no evidence, she said. "Part of healthcare reform has to look at the decision-making process of who gets what, when, and not just whether it gets paid for or not."

One participant noted that creating a prospective registry of test consumers—primarily practitioners—could indicate which predictive tests are being used by doctors for which purposes, and what clinical decisions are being influenced by the test results. Although a registry would not be as robust as a clinical trial, it could still provide some useful information. Another participant pointed out that the registry of treatment outcomes created by the Cystic Fibrosis Foundation, which includes data from 150 cystic fibrosis centers throughout the country, has led to dramatic improvements in treatments and outcomes for cystic fibrosis (CFF, 2009). "It led to a tremendous dialogue between patients and centers as to what they are actually doing, in terms of quality care, which has now led to many centers improving their performance," he said. Similar non-regulatory strategies for collecting and improving the clinical utility of predictive tests could be implemented. Ms. Stack noted that many predictive tests, such as Oncotype DX, provide risk information, but ultimately, doctors make the final call on what that risk means—they determine their own cutoff points for risk that warrants treatment or not. "Ultimately, the doctor's going to make that treatment call and have a lot of data about it," she said, so a registry that collects that data would be useful in ascertaining the clinical utility of a test.

Dr. Ferreira-Gonzales noted the catch-22-like nature of clinical utility determinations. "Some third-party payers are making decisions on the lack of information that we have in these areas. But if we don't offer the testing and know how it was actually being used, we will never know this information. So we need to be able to gather this information on what the test does for the patients, and SACGHS discussed the possibility of making a decision to allow a test to enter the market dependent on evidence development via a registry, as has been done with CT scans and in other areas of medicine." Dr. Burke added that "there is a need to know, not only what we know, but what we don't know, because it's what we don't know that points to the critical research that needs to be done. Clinicians on the front lines are a very good source for that kind of information."

A participant stressed the need for a test registry to report not only positive results of studies, but also negative results. "Another registry concept would be the registration of validation studies prior to their initiation so that we could follow up and make sure that the results of those studies are subsequently presented and published." Dr. Ferreira-Gonzalez responded by pointing out that one of SACGHS's recommendations was that both positive and negative results should be shared in a Web-based system (SACGHS, 2008b).

Dr. Friend agreed that the experimental evidence should be posted in registries and made available to the public. However, he cautioned against having the FDA be responsible for the registry. "I'm not sure the FDA needs to do that," he said. The FDA should be responsible for regulating analytical and clinical validity, but not clinical utility. Mr. Erwin added, "When I hear about proposals that the FDA should regulate everything, on the one hand you don't want to see innovation delayed or stifled. On the other hand, if the standards are not high, the promise of personalized medicine will never be realized, because there will be no real incentive to put the money and time that's necessary to do it right in order to get rewarded financially or professionally. We need rigor without rigidity. As technology evolves and as unexpected things come along, the regulatory framework has to be adaptable enough to deal with that without extremely long delays. If new technology can't fit into a box that's so rigid that we can't derive the benefit from it, then there's something wrong with regulation."

Dr. Parkinson concurred, calling for more flexible regulation of predictive tests. "These tests are going to have to continue to evolve as the therapeutics are evolving, and it's almost like an iterative process. There needs to be informed regulation. This requires a tighter link between biological characterization, and predictive test development and therapeutic applications—a strategic approach to the regulation recognizing that these diseases are being redefined by the tests, and by the effect of therapeutics on patients characterized by these tests." Dr. Parkinson called for having some ongoing review of new information, akin to what EGAPP does.

Dr. Mansfield cautioned against just putting tests on the market before adequately assessing their safety and effectiveness, and relying on a registry to determine clinical utility. "There is already a mechanism for tests to go to market before we know all their performance mechanisms, and it's called an investigational device exemption. It seems to work very well," she noted.

Regardless of how clinical utility is assessed, it is a costly endeavor that needs more federal financial support. Dr. Burke noted that SACGT asked all federal agencies to indicate how much work they did in research relevant to evaluating tests, and found "there was tremendous room for growth in federal funding of research around the assessment of clinical utility. We noted that healthcare funding decisions often function as an oversight mechanism." SACGT also recommended more federal government support for evidence-based guideline development related to predictive tests. This recommendation has been realized to some degree through the U.S. Preventive Services Task Force (USPSTF) and EGAPP reviews of predictive tests, Dr. Burke noted.

Reimbursement

In addition to technological and regulatory hurdles to the development of personalized medicine in oncology, reimbursement hurdles also exist. Drs. Jeff Roche and Amy Bassano of CMS described how Medicare decides what predictive tests to reimburse and how much to reimburse for a test. Their presentations were followed by critiques of the current reimbursement system, as well as suggestions for improving the reimbursement of predictive tests.

MEDICARE COVERAGE OF PREDICTIVE TESTS

Dr. Roche explained that the Social Security Act of 1965 established Medicare, a health insurance program run by the U.S. government for individuals age 65 and over, or for individuals who meet special criteria.[1] Medicare pays for services and items that are reasonable and necessary for the diagnosis or treatment of illness or injury in those who qualify for the program.[2] In general, Medicare is not required to cover screening services, although there are certain exceptions (e.g., Medicare covers the cost of Pap tests, colorectal cancer screening tests, mammograms, and the PSA screening test). The diagnostic services that Medicare covers are done in a variety

---

[1] The Social Security Act of 1965. Public Law 89-97. (July 30, 1965).
[2] The Social Security Act. 42 U.S.C. § 1862(a)(1)(A) (2009).

of settings, Dr. Bassano noted, including hospitals, physician offices, and independent labs.

For a medical intervention to qualify as reasonable and necessary, evidence must show, among other considerations, that the item or service improves clinically meaningful health outcomes in Medicare beneficiaries (CMS, 2009c). Evidence is assessed using standard principles of evidence-based medicine. CMS generally follows the evaluation process developed by other agencies or advisory bodies, such as AHRQ, USPSTF, and EGAPP. Genetic test coverage determinations are particularly challenging because genetic tests can be used for both diagnostic and screening purposes, Dr. Roche said. In addition, the evidence base is small for genetic tests, and the science is evolving. "The ultimate health outcomes attributable to genomic testing are not clear at this time," Dr. Roche said, and there can be dangers in making some coverage determinations prematurely.

More recently, CMS has begun using criteria for Analytic validity; Clinical validity; Clinical utility; and Ethical, legal, and social implications (ACCE) (CDC, 2009) in making decisions on whether sufficient evidence exists to justify coverage of predictive tests. However, CMS has not yet formally adopted the ACCE framework or any other framework of evidence for predictive testing. In general, CMS rates evidence according to health outcomes. Diagnostic tests that lead to longer life expectancy, improved function, or significant symptom improvement are rated higher than tests that result in doctor confidence or earlier detection without improved survival (Box 1).

Some Medicare coverage decisions are made at the national level of the organization, but approximately 85 to 90 percent of coverage decisions are made by local contractors (CMS, 2009b). Local contractors can increase national coverage and reimburse additional procedures. For example, some local contractors cover gene marker tests for hereditary cancer syndromes, including ovarian cancer and colorectal cancer, assuming certain conditions are met (i.e., the individual is clinically affected by the disorder and is willing to undergo pretest genetic counseling). While some contractors do not cover these gene marker tests, local coverage of genetic analyses must be provided through a laboratory that meets ASCO's recommended requirements, and the patient must sign an informed consent form prior to testing.

Dr. Ratain pointed out that because laboratories can receive samples for testing from all over the country, some local coverage decisions have national ramifications. For example, a California Medicare contractor made

---

BOX 1
Rating Evidence of Health Outcomes

More Impressive
- Longer life and improved function/participation
- Longer life with arrested decline
- Significant symptom improvement allowing better function/ participation
- Reduced need for further burdensome tests and treatments

Less Impressive
- Earlier detection without improved survival
- Test result is a better number
- Image/scan looks better
- Doctor feels more confident

SOURCE: CMS presentation (June 9, 2009).

---

a decision in 2007 to cover chemotherapy sensitivity testing, and laboratories that provide this testing in southern California can receive samples from anywhere in the United States. Dr. Ratain believes there is inherent unfairness in a system that enables a laboratory in one area of the country where the predictive test is covered to do nationwide testing, yet deny reimbursement for such testing to a laboratory located in other areas with different local coverage determinations. Dr. Roche acknowledged the inconsistency in policy, but pointed out that "when you lose local coverage discretion, you're also losing the ability of a local coverage organization to respond to the needs that are being expressed in that region of the country."

Dr. Roche also stated that CMS often consults with experts in other government agencies, such as those at the FDA and CDC, as well as with professional societies such as ASCO, the American Society for Hematology, and the American College of Chest Physicians, when making coverage decisions. "We are now realizing that things that are embedded outside of CMS deserve more than a second look, and we're trying to integrate this information and understand better some of the implications [of coverage decisions] on the system," he said.

REIMBURSEMENT RATES

Dr. Bassano explained how Medicare determines the payment rate for diagnostic tests. The Medicare payment rate for diagnostics is calculated as the lesser of

- the amount billed;
- the local fee for the area; or
- the national limitation amount (NLA) for the particular code.

The NLAs are on a fixed-fee schedule. For tests that had NLAs established before January 1, 2001, the NLA is 74 percent of the median of all local fee schedule amounts. NLAs established on or after January 1, 2001, are 100 percent of the median. Fees may be updated by statute, but are not updated regularly. No updates occurred between 2004 and 2008. In 2009, fee rates were increased by 4.5 percent. "This is an older system that hasn't had the same type of updating or scrutiny by Congress that some of the other physician or hospital systems have. It's not resource based, nor is it a prospective payment system, and there's really no opportunity for CMS to reassess the payment rates for the codes," said Dr. Bassano. Furthermore, there are no budget neutrality requirements for Medicare reimbursement of diagnostics.

Dr. Quinn asserted that the fixed-fee schedule for diagnostics hampers innovation. For example, fee rates are currently around $14 for all PSA tests, he said. This test frequently has been criticized by clinicians for insufficient specificity. If an improved version of the test was developed, this would require substantial resources. However, the new test would receive reimbursement at the same rate as the older, less effective PSA test because of the lack of specificity in the fee schedule.

There is a process, however, for annually updating the Current Procedural Terminology (CPT) codes for new diagnostics. (The CPT code categorizes the diagnostic and ultimately determines its fee rate.) To determine the CPT code for new tests, the code can be crosswalked to an existing test, pricing the new test at the same rate as a similar test that already has a CPT code, or "gap filled." Gap filling requires Medicare contractors to collect data specific to their geographic area and to set a new price that reflects those data. This process is burdensome, so most new code pricing is determined by crosswalking. CMS collects public feedback on the new code pricing before putting it into practice.

Dr. Bassano noted that many laboratory-developed tests come from independent labs that are not required to have a CPT code. Instead they are given an unlisted code that the lab can use to bill, and that price is set by the local contractor. These tests tend to be more expensive, costing as much as thousands of dollars. Also, this coding process is not tracked as well as that for tests using standard CPT codes, Dr. Bassano pointed out.

Alternatively, laboratory-developed tests can use a "stack" of generic chemical-test steps represented by CPT codes such as "83898 DNA Amplification" (AMA, 2009). These stacked CPT codes are often difficult to track, Dr. Quinn said. "You get a list of 30 CPT codes, and have no idea what the test is, what was done, what the accuracy is, what the characteristics are, and what the utility is." Stacked codes can also provide disincentives to develop step-saving innovations, Dr. Shak added, because the current system of CPT codes pays for activity, not value. For example, an older test to detect methylation known as the Southern methylation analysis requires six steps, whereas a newer PCR methylation test requires only four of those steps (Table 2). Consequently, the older test is reimbursed at a higher rate, despite the fact that the newer test has improved dependability and performance, eliminates the need for radioactivity, and produces faster results. Laboratories that choose to do the better PCR-based methylation test are paid less than those that

TABLE 2 Current System of CPT Codes Pays for Activity, Not Value

	CPT Code	Process	Units	Rates
Southern methylation analysis	1	Extraction	1	$26.00
	2	Digestion	1	$51.00
	3	Separation	1	$26.00
	4	Nucleic acid probe	1	$26.00
	5	Southern blot	1	$52.00
	6	Interpretation/report	1	$26.00
			Total =	$207.00
PCR methylation analysis	1	Extraction	1	$26.00
	2	Digestion	1	$51.00
	3	Separation	1	$26.00
	4	Interpretation/report	1	$26.00
			Total =	$129.00

NOTE: CPT = Current Procedural Terminology, PCR = polymerase chain reaction.
SOURCE: Shak (2009).

continue to use the older test. "The rewarding of activity perversely can lead to the performance of lots of unnecessary steps," Dr. Shak said.

Several presenters and speakers said another major problem with the reimbursement rates for predictive tests is that they are far lower than therapeutics. For example, the Oncotype DX test, which identifies node-negative, ER-positive breast cancer patients for whom chemotherapy is unlikely to help, costs $3,500 a test, Dr. Hayes said. Yet he estimates that the test provides a net healthcare savings of about $1 billion annually by preventing the ineffective use of chemotherapy. Most health insurers balk at the prospect of paying $3,500 for a test, but "that's cheap if it's going to save $50,000 in chemotherapy expenses," he said. Dr. Quinn added, "Cost-saving innovations are potentially huge. We just need a pathway to make it possible."

Additionally, the high complexity of the technology used in many predictive tests, such as gene splicing, should drive up the cost of the tests, Dr. Quinn stressed. "Every test is almost like a master's degree back when I was in college." He estimated that developing a new test can cost more than $50 million, but may only be used by 10,000 patients a year. Breaking even on development costs in 5 years would require recouping about $10 million, which boils down to a fee of a few thousand dollars per test. This level of reimbursement would not result in a profit or cover operating costs and other company expenses linked to the test, Dr. Quinn noted. Ms. Stack concurred, noting that "It's expensive to do quality development and we need to be able to charge a fair market value [for our tests], like we do for drugs. It's not too much to ask to charge 10 to 20 percent of what a therapy would cost for a diagnostic that's innovative."

BUNDLING OF PAYMENTS

Dr. Bassano stated that Medicare prefers to have one payment for all the services provided to a patient during a hospital stay. This often means that the reimbursement for a test done on a specimen collected during a hospital stay is bundled with a reimbursement payment for other hospital services. The bundled payment is made to the hospital, not to the laboratory doing the testing. Since 2001, Medicare rules state that the date of service for reimbursed laboratory services is generally the date the specimen is collected (CMS, 2009a). If a specimen is stored less than 30 days, payment for the test performed on that specimen is bundled into the payment for the inpatient hospital stay, and there is no separate payment for the test,

Dr. Bassano explained (CMS, 2009a). An outpatient center can also bill for the test. For tests done on specimens stored for more than 30 days, the date of service is the date the specimen was removed from storage, and payment is separate from inpatient payments (CMS, 2009a). In this case, the lab doing the test does the billing, rather than a hospital.

In response to complaints by stakeholders, this rule was modified in 2007 to allow some payments for tests to be unbundled from the payment for hospital service. This exception applies if the test was performed on a specimen collected during a hospital stay and stored for less than or equal to 30 days, and one of the following conditions applies:

- The test was ordered at least 14 days following the date of the patient's discharge from the hospital.
- The specimen was collected while the patient was undergoing a hospital surgical procedure.
- It would be medically inappropriate to have collected the sample other than during the hospital procedure for which the patient was admitted.
- The results of the test do not guide treatment provided during the hospital stay.
- The test was reasonable and medically necessary for treatment of an illness.

These exceptions are viewed as insufficient by some stakeholders, Dr. Bassano noted. She said industry claims the inpatient bundling of payments was not intended to cover the costs of expensive, complex tests, and that such bundling inhibits the development of tests performed in a single location. Laboratories do not want to negotiate the reimbursement rates for their tests with hospitals located throughout the country, and would rather negotiate directly with local Medicare contractors, Dr. Bassano said.

Dr. Quinn concurred, saying, "This means that the lab has to contract with 5,000 hospitals, instead of one Medicare program or contractor. So the transaction costs go through the roof." Hospitals also do not want to take on the audit risk, he added; there is an auditing requirement that hospitals track down all the hospital and doctor records involved in the testing done on an oncology specimen acquired at the hospital. Often, specimens are transported to multiple physicians or cancer facilities throughout the country, so this is a major undertaking. In addition, by bundling tests into hospital payments, CMS loses the opportunity to recoup payment if its audit of a lab determines the service was not medically necessary. "Once the test is being done all across the country, in 3,000 hospitals, and with an unlisted code, it's no longer auditable," Dr. Quinn said.

Ms. Stack observed that "a lot of our companies are testing samples that were collected in a hospital, and they don't want to wait 15 days to do the genomic tests so that they can be in accordance with the 14-day rule," she said. Another problem with the rule is that it creates a bias against inpatient cancer care because such care requires bundling of the test reimbursement into the total hospital payment, rather than reimbursing the lab directly, Dr. Quinn said. "If you have something like a brain tumor that requires inpatient surgery, you'd be biased against developing a test for it, but if it's a lumpectomy with a lot of outpatient surgery, you'd be more biased to develop a test for it. Large companies are very aware of that," he said.

Dr. Quinn also questioned the logic of how Medicare bundles its payments, and called for more economically rational bundling. "If you bundled the cost of a $2,000 lab test backward to an $80 office visit or a $4 blood draw, it's very hard to make economic sense out of that. If you would bundle that cost forward to the $50,000 chemotherapy, people would be knocking over themselves to use that test. So I think forward bundling could make a lot of sense if it was tied to chemotherapy. There are old rules that don't apply now, and have really detrimental effects on the development of this [genetic testing] industry," he said

VALUE OF BIOMARKERS

Dr. Hayes called for valuing markers as much as we value therapeutics. This will "require a wholesale change of the system," he said. "A bad tumor marker is as harmful as a bad drug," yet the evidence of safety and effectiveness required to put a tumor marker on the market is far lower than that for a tumor therapeutic. "We would not let drugs into the market based on Phase I data," he said. "We insist on showing efficacy and safety before we allow people to use the drug, and we don't just say, 'well let's get it out there and see if people like it' because I think we convince or trick ourselves into thinking we like something when, in fact, it may not be helpful."

The lack of evidence on tumor markers is appalling, Dr. Hayes stressed. Over the past 14 years, the ASCO Tumor Marker Guidelines Panel has only recommended the use of four tumor markers, despite publications on hundreds of such putative markers. The other markers the panel reviewed lacked sufficient evidence. Most tumor marker studies are tested in retrospective trials with multivariate or univariate analyses, he said, and have small sample sizes. Markers are not often tested in sufficiently powered studies or meta-analysis studies in which the marker was the primary objective of prospective testing, or in prospective studies in which the marker was the secondary objective (Hayes et al., 1996).

Despite their insufficiencies, the results of many tumor marker studies are published in reputable scientific journals, Dr. Hayes said. However, it takes years for this type of study to gather enough evidence on clinical utility. "We could truncate this entire process considerably by looking at markers the way we look at drugs," he said. The clinical utility of a tumor marker can be shown through just one well-designed trial, he said, but using archived samples with many built-in biases could take two or three trials. "This makes it harder up front, but in the long run you actually get answers quicker that way," Dr. Hayes said.

Unfortunately, doing large, controlled, prospective clinical trials of tumor markers would be much more costly, and untenable given the current reimbursement rates for predictive tests, Dr. Hayes said. "If we increase the rigor required to introduce a new predictive tests to that required to introduce a new drug, current reimbursement systems will smother innovation," he said. Consequently, a vicious cycle is created: Because predictive tests are insufficiently reimbursed, shortcuts are taken in their development to save money, and their clinical safety and effectiveness is not adequately determined. As a result of their questionable utility, insurers continue to undervalue them and thus provide inadequate reimbursement (Figure 8). Given the low rates at which tests are reimbursed, and the inconsistent and often minimal amount of regulatory oversight on tests, "there is little incentive to do properly designed and controlled clinical trials," Dr. Hayes said. "Therefore, there is a much lower level of evidence for markers than there is for drugs, less certainty of data, and less value for tumor marker clinical utility because people don't know how to use them."

In addition, the NIH, academic institutions, and other sponsors of clinical research do not value biomarkers as much as they do drugs, and consequently do not provide adequate support for clinical trials of these markers, Dr. Hayes said. This adds to the vicious cycle. "Tumor marker research is not perceived to be as exciting or as important as new therapeutics, especially the clinical component. There is less academic credit and funding for those of us who do tumor marker work." Dr. Leonard concurred, saying, "The only thing we have to work with are convenient samples because there is no funding. That's an NIH decision. So we have

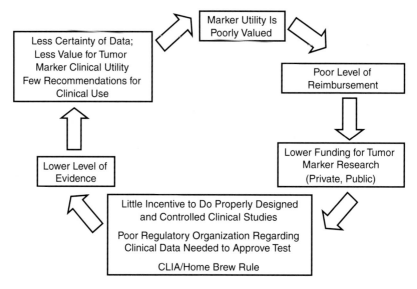

FIGURE 8 Undervalue of tumor markers: A vicious cycle.
NOTE: CLIA = Clinical Laboratory Improvement Amendments.
SOURCE: Hayes Presentation (June 9, 2009).

to [work] with in vitro diagnostic companies that don't pay, and that's not respected in academia. It's a terrible system and we have to fix it."

Dr. Hayes proposed what he called a "virtuous cycle" to replace the current vicious cycle (Figure 9). In the virtuous cycle, tumor markers would be highly valued, and thus researchers would receive greater funding for clinical trials to assess them. This would result in better evidence of their utility and require higher levels of reimbursement. The end result would be strong recommendations for clinical use of tumor markers that would occur much sooner, with proven efficacy. The study section the NCI recently started for cancer biomarker research is a first step toward implementing the virtuous cycle, Dr. Hayes noted. Previously, grants for clinical research on tumor markers were inappropriately reviewed by the pathology or therapeutic study sections. Another positive step is the fact that criteria for reporting tumor marker studies have been recently developed and adopted by scientific journals (Bossuyt et al., 2004; McShane et al., 2005).

To achieve the virtuous cycle, Dr. Hayes called on the patient advocacy community to promote the value of tumor markers, and to insist that CMS provide a higher level of reimbursement for predictive tests, commensurate

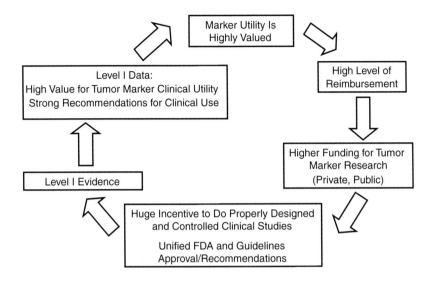

FIGURE 9 Highly valued tumor markers: Virtuous cycle.
NOTE: FDA = Food and Drug Administration.
SOURCE: Hayes Presentation (June 9, 2009).

with what is provided for drugs and with a more rigorous approval process for tests. Dr. Phillips added that private payers are key drivers in the reimbursement system and need to change their reimbursement rates for predictive tests.

In addition, Dr. Hayes suggested changing the method of caregiver reimbursement so that doctors can spend more time with their patients explaining predictive tests, and not be financially penalized for not recommending chemotherapy because the test indicates it will not be effective. "I can spend 15 minutes with a patient and say she should get chemotherapy, and she'll probably be quite happy because it sounds like I'm being aggressive and doing the right thing, or I can spend 45 minutes to an hour explaining what the 21-gene recurrence score is, and why that patient probably won't benefit from chemotherapy. If I do the latter, I think I'll get $220 for the visit. If I do the former, my institution and I get about several thousand dollars. That's not right. It should be the same, and I think we need to figure out how to make it the same," Dr. Hayes said.

"Paying for outcomes will [also] drive the development of markers, and until we pay for outcomes, the likelihood that there's going to be money to develop a predictive test is going to be small," said Dr. Friend. Dr. Parkinson added, "Maybe low levels of evidence will get low levels of reimbursement, and high levels of evidence and clinical utility will get higher levels of reimbursement. So all of a sudden, the system will be motivated to perform. We're starting to see pay-for-performance on the therapeutics development side. Maybe it's time to have that on the diagnostics development side." Dr. Ratain said it might be worthwhile to "create a completely level playing field where all medical technologies are reimbursed as something that is somehow tied to value." Basing reimbursement rates for therapeutics on the value they provide might lower the cost of drugs and provide funds for increasing the price of predictive tests.

Several participants called for early reimbursement of predictive tests. Dr. McCormick of Veridex noted that small device companies such as his own operate differently from drug companies because they have less financial assets to do extensive clinical trials. Early reimbursement helps a device manufacturer to fund these trials. Dr. Herbst also urged early reimbursement in clinical trials of tests that assess multiple tumor markers "so you can [assess] the mutations you're interested in and then look at others." Both Drs. Johnson and Small noted that it can be problematic to run clinical trials of predictive tests because it is questionable whether health insurers will reimburse the cost of the tests.

Dr. Friend suggested developing an accelerated approval process for the codevelopment of a diagnostic and therapeutic. Alternatively, a predictive test could reach the market quickly for a new indication through off-label use. However, Dr. Quinn noted that off-label use of a diagnostic by the company that makes the test creates a conflict of interest, which is different from that seen in off-label use of drugs or devices. "We have this conflict that is more direct for innovative lab tests because the same company that produces the test—does the innovation—is also selling it, and that doesn't occur anywhere else. GE makes positron emission tomography (PET) scanners and spends a billion dollars developing a new PET scanner. But then it sells them, and a hospital and a doctor will get the profit or manage the use," he said. To recoup their innovation costs, companies that make laboratory-developed tests will be inclined to overpromote and overuse their tests.

Dr. Quinn also criticized relying on evidence-based medicine to determine use and reimbursement of tests. "Say I have colon cancer with a 10 percent chance of recurrence, and I have a PET scan that shows golf ball–sized lesions all over my innards. This test has an odds ratio of 3, which means I have a 30 percent chance of having recurrent colon cancer. It doesn't make any sense in the context of the diagnostic test," he said, adding, "We have no good standard for coverage decisions. We say, 'you need more evidence,' but there's no unit of evidence. You don't measure evidence in cubic feet or in meters. We say 'you need more evidence,' but against what standard? There's absolutely none." He stressed that predictive tests are conceptually quite different from therapeutics, and should be evaluated differently.

Summary

During 2 lively days of discussion, it became apparent that predictive tests that enable cancer treatments to be tailored to the highly specific biochemical abnormalities that underlie a tumor, rather than to the more general pathology, hold great promise for making cancer therapy more safe and effective. These tests can be more predictive of treatment response than standard clinical prognostics, and have already become part of standard clinical practice for some cancers and cancer drugs. However, in order to realize the promise of personalized cancer medicine, a number of obstacles need to be overcome, including technological, regulatory, and reimbursement hurdles.

On the technological side, speakers and participants noted that the research community needs to improve its understanding of genetic pathways and how predictive tests work. It also needs to develop superior methods of predictive test validation, improve test reliability, and advance how predictive tests are used in clinical decision making. In addition, methods for codeveloping biomarkers concurrently with targeted drugs need improvement.

Many workshop participants expressed concern about the disparities in the regulation of laboratory-developed tests and FDA-approved tests. This lack of a well-defined process for biomarker development, validation, qualification, and use has reduced interest and investment in developing predictive tests. Speakers and participants suggested that the regulatory system needs to be dynamic and able to adapt to rapid changes in technology.

Finally, speakers and participants proposed that the reimbursement system needs to be adjusted to reward the development and use of high-quality predictive tests. The current reimbursement system of coding, bundling of payments, and using a fixed-fee schedule for predictive tests discourages test innovation; does not adequately recognize the clinical importance of predictive tests; and is not value based. An IOM committee will examine these issues further, and develop consensus-based recommendations for moving the field of personalized medicine forward.

References

AHRQ (Agency for Healthcare Research and Quality). 2008. *Impact of gene expression profiling tests on breast cancer outcomes.* Rockville, MD: AHRQ.

AMA (American Medical Association). 2009. *CPT-current procedural terminology.* http:// www.ama-assn.org/ama/pub/physician-resources/solutions-managing-your-practice/ coding-billing-insurance/cpt.shtml (accessed October 5, 2009).

Amgen. 2008. *Prescribing information.* http://www.vectibix.com/prescribing_information/ prescribing_information.html (accessed September 29, 2009).

Andriole, G. L., E. D. Crawford, R. L. Grubb, III, S. S. Buys, D. Chia, T. R. Church, M. N. Fouad, E. P. Gelmann, P. A. Kvale, D. J. Reding, J. L. Weissfeld, L. A. Yokochi, B. O'Brien, J. D. Clapp, J. M. Rathmell, T. L. Riley, R. B. Hayes, B. S. Kramer, G. Izmirlian, A. B. Miller, P. F. Pinsky, P. C. Prorok, J. K. Gohagan, C. D. Berg, and P. P. T. the. 2009. Mortality Results from a Randomized Prostate-Cancer Screening Trial. *New England Journal of Medicine* 360(13):1310–1319.

AstraZeneca. 2009. *Iressa (gefitinib) recommended for approval for the treatment of non-small cell lung cancer in Europe* http://www.astrazeneca.com/media/latest-press-releases/2009/ iressa-chmp?itemId=5585247 (accessed September 8, 2009).

Bild, A. H., G. Yao, J. T. Chang, Q. Wang, A. Potti, D. Chasse, M. B. Joshi, D. Harpole, J. M. Lancaster, A. Berchuck, J. A. Olson, Jr., J. R. Marks, H. K. Dressman, M. West, and J. R. Nevins. 2006. Oncogenic pathway signatures in human cancers as a guide to targeted therapies. *Nature* 439(7074):353–357.

Blum, R., R. Elkon, S. Yaari, A. Zundelevich, J. Jacob-Hirsch, G. Rechavi, R. Shamir, and Y. Kloog. 2007. Gene expression signature of human cancer cell lines treated with the Ras inhibitor salirasib (*S*-Farnesylthiosalicylic acid). *Cancer Research* 67(7):3320–3328.

Bossuyt, P. M., J. B. Reitsma, D. E. Bruns, C. A. Gatsonis, P. P. Glasziou, L. M. Irwig, J. G. Lijmer, D. Moher, D. Rennie, and H. C. W. de Vet. 2004. Towards complete and accurate reporting of studies of diagnostic accuracy: The STARD initiative. *Family Practice* 21(1):4–10.

CAP (College of American Pathologists). 2007. *HER2 and you: Guidelines provided by CAP and ASCO.* Northfield, IL: CAP.

Cartron, G., L. Dacheux, G. Salles, P. Solal-Celigny, P. Bardos, P. Colombat, and H. Watier. 2002. Therapeutic activity of humanized anti-CD20 monoclonal antibody and polymorphism in IgG Fc receptor Fc RIIIa gene. *Blood* 99(3):754–758.

CDC (Centers for Disease Control and Prevention). 2009. *Genomic translation: ACCE model process for evaluating genetic tests.* http://www.cdc.gov/genomics/gtesting/ACCE/index. htm (accessed September 10, 2009).

CFF (Cystic Fibrosis Foundation). 2009. *Patient registry report.* http://www.cff.org/research/ ClinicalResearch/PatientRegistryReport/ (accessed September 28, 2009).

CMS (Centers for Medicare & Medicaid Services). 2009a. *Medicare claims processing manual. Chap. 16, laboratory services.* http://www.cms.hhs.gov/manuals/Downloads/clm104c16. pdf (accessed September 17, 2009).

CMS. 2009b. *Medicare coverage database.* http://www.cms.hhs.gov/mcd/indexes.asp (accessed September 28, 2009).

CMS. 2009c. *Proposed decision memo for pharmacogenomic testing for warfarin response (CAG-00400N).* http://www.cms.hhs.gov/mcd/viewdraftdecisionmemo.asp?from2=viewdraft decisionmemo.asp&id=224& (accessed September 24, 2009).

Di Nicolantonio, F., M. Martini, F. Molinari, A. Sartore-Bianchi, S. Arena, P. Saletti, S. De Dosso, L. Mazzucchelli, M. Frattini, S. Siena, and A. Bardelli. 2008. Wild-type BRAC is required for response to panitumumab or cetuximab in metastatic colorectal cancer. *Journal of Clinical Oncology* 26(35):5705–5712.

EGAPP Working Group (Evaluation of Genomic Applications in Practice and Prevention Working Group). 2009a. Recommendations from the EGAPP Working Group: Can tumor gene expression profiling improve outcomes in patients with breast cancer? *Genetics in Medicine* 11(1):66–73.

EGAPP Working Group. 2009b. Recommendations from the EGAPP Working Group: Can UGT1A1 genotyping reduce morbidity and mortality in patients with metastatic colorectal cancer treated with irinotecan? *Genetics in Medicine* 11(1):15–20.

FDA (Food and Drug Administration). 1997. Medical devices; classification/reclassification; restricted devices; analyte specific reagents—FDA. Final Rule. *Federal Register* 62(225):62243–62260.

FDA. 2005. *Drug-diagnostic co-development concept paper.* Silver Spring, MD: FDA.

FDA. 2007a. Commercially Distributed Analyte Specific Reagents (ASRs): Frequently asked questions. http://www.cytometry.org/website_pages/FDA%20ASR%20rule%202007. pdf (accessed September 19, 2009).

FDA. 2007b. *Draft guidance for industry, clinical laboratories, and FDA staff: In vitro diagnostic multivariate index assays.* Rockville, MD: FDA.

FDA. 2009a. *Intercenter agreement between the Center for Drug Evaluation and Research and the Center for Devices and Radiological Health.* http://www.fda.gov/CombinationProducts/ JurisdictionalInformation/ucm121177.htm (accessed September 28, 2009).

FDA. 2009b. *Medical devices: General and special controls.* http://www.fda.gov/MedicalDevices/ DeviceRegulationandGuidance/Overview/GeneralandSpecialControls/default.htm (accessed September 23, 2009).

FDA. 2009c. *Medical devices: In vitro diagnostics.* http://www.fda.gov/MedicalDevices/ ProductsandMedicalProcedures/InVitroDiagnostics/default.htm (accessed September 23, 2009).

FDA. 2009d. *Medical devices: Postmarket requirements (devices).* http://www.fda.gov/ MedicalDevices/DeviceRegulationandGuidance/PostmarketRequirements/default.htm (accessed September 23, 2009).

Friend, S. H. 2009. *Multiple biomarkers efficiently identify equivalent patient populations (which marker to use?/how will we know?).* PowerPoint presentation, June 8. Washington, DC: Merck Research Laboratories.

Genentech. 2008. Genentech, Inc. Citizen Petition: Regulation of *In Vitro* Diagnostic Tests. http://www.regulations.gov/search/Regs/home.html#documentDetail?R=0900006480 7d4a7e (accessed December 5, 2008).

Genomic Health. 2009. *Oncotype DX® guides individualized treatment decisions for more than 100,000 breast cancer patients worldwide.* http://investor.genomichealth.com/ releasedetail.cfm?ReleaseID=386285 (accessed September 3, 2009).

Hayes, D. F., R. C. Bast, C. E. Desch, H. Fritsche, Jr., N. E. Kemeny, J. M. Jessup, G. Y. Locker, J. S. MacDonald, R. G. Mennel, L. Norton, P. Ravdin, S. Taube, and R. J. Winn. 1996. Tumor marker utility grading system: a framework to evaluate clinical utility of tumor markers.[see comment]. *Journal of the National Cancer Institute* 88(20):1456–1466.

ImClone Systems. 2008. *Highlights of prescribing information: Erbitux.* http://packageinserts. bms.com/pi/pi_erbitux.pdf (accessed September 29, 2009).

IOM (Institute of Medicine). 2007. *Cancer biomarkers: The promises and challenges of improving detection and treatment.* Washington, DC: The National Academies Press.

Jonker, D. J., C. J. O'Callaghan, C. S. Karapetis, J. R. Zalcberg, D. Tu, H.-J. Au, S. R. Berry, M. Krahn, T. Price, R. J. Simes, N. C. Tebbutt, G. van Hazel, R. Wierzbicki, C. Langer, and M. J. Moore. 2007. Cetuximab for the treatment of colorectal cancer. *New England Journal of Medicine* 357(20):2040–2048.

Jonker, D. J., C. Karapetis, C. Harbison, C. J. O'Callaghan, D. Tu, R. J. Simes, L. Xu, M. J. Moore, J. R. Zalcberg, and S. Khambata-Ford. 2009. High epiregulin (EREG) gene expression plus KRAS wild-type (WT) status as predictors of cetuximab benefit in the treatment of advanced colorectal cancer (ACRC): Results from NCIC CTG CO.17—A Phase III trial of cetuximab versus best supportive care (BSC). *Journal of Clinical Oncology (Meeting Abstracts)* 27(15S):4016.

Khoury, M. J., M. Gwinn, P. W. Yoon, N. Dowling, C. A. Moore, and L. Bradley. 2007. The continuum of translation research in genomic medicine: How can we accelerate the appropriate integration of human genome discoveries into health care and disease prevention? *Genetics in Medicine* 9(10):665–674.

Khoury, M. J., W. G. Feero, M. Reyes, T. Citrin, A. Freedman, D. Leonard, W. Burke, R. Coates, R. T. Croyle, K. Edwards, S. Kardia, C. McBride, T. Manolio, G. Randhawa, R. Rasooly, J. St. Pierre, and S. Terry. 2009a. The Genomic Applications in Practice and Prevention Network. *Genetics in Medicine* 11(7):488–494.

Khoury, M. J., C. M. McBride, S. D. Schully, J. P. A. Ioannidis, W. G. Feero, A. C. J. W. Janssens, M. Gwinn, D. G. Simons-Morton, J. M. Bernhardt, M. Cargill, S. J. Chanock, G. M. Church, R. J. Coates, F. S. Collins, R. T. Croyle, B. R. Davis, G. J. Downing, A. Duross, S. Friedman, M. H. Gail, G. S. Ginsburg, R. C. Green, M. H. Greene, P. Greenland, J. R. Gulcher, A. Hsu, K. L. Hudson, S. L. R. Kardia, P. L. Kimmel, M. S. Lauer, A. M. Miller, K. Offit, D. F. Ransohoff, J. S. Roberts, R. S. Rasooly, K. Stefansson, S. F. Terry, S. M. Teutsch, A. Trepanier, K. L. Wanke, J. S. Witte, and J. Xu. 2009b. The scientific foundation for personal genomics: Recommendations from a National Institutes of Health–Centers for Disease Control and Prevention multidisciplinary workshop. *Genetics in Medicine* 11(8):559–567.

Levis, M., P. Brown, B. D. Smith, A. Stine, R. Pham, R. Stone, D. DeAngelo, I. Galinsky, F. Giles, E. Estey, H. Kantarjian, P. Cohen, Y. Wang, J. Roesel, J. E. Karp, and D. Small. 2006. Plasma inhibitory activity (PIA): a pharmacodynamic assay reveals insights into the basis for cytotoxic response to FLT3 inhibitors. *Blood* 108(10):3477–3483.

Love/Avon Army of Women. 2009. *About us.* http://www.armyofwomen.org/aboutus (accessed September 10, 2009).

Mass, R. 2009. *Technology hurdles: Drug developer perspective.* PowerPoint presentation, June 8. Washington, DC: Genentech, Inc.

McShane, L. M., D. G. Altman, W. Suerbrei, S. E. Taube, M. Gion, G. M. Clark, and for the Statistics Subcommittee of the NCI—EORTC Working Group on Cancer Diagnostics. 2005. Reporting recommendations for tumor MARKer prognostic studies (Remark). *Nature Clinical Practice Oncology* 2(8):416–422.

Meshinchi, S., T. A. Alonzo, D. L. Stirewalt, M. Zwaan, M. Zimmerman, D. Reinhardt, G. J. Kaspers, N. A. Heerema, R. Gerbing, B. J. Lange, and J. P. Radich. 2006. Clinical implications of FLT3 mutations in pediatric AML. *Blood* 108(12):3654–3661.

NCI (National Cancer Institute). 2008. *Personal genomics: Establishing the scientific foundation for using personal genome profiles for risk assessment, health promotion, and disease prevention.* Bethesda, MD: NCI.

NCI. 2009a. *NCI guidelines for ARRA research and research infrastructure grand opportunities: Comparative effectiveness research in genomic and personalized medicine: Announcement number: RFA-OD-09-004.* http://www.cancer.gov/pdf/recovery/004_cer_personalized_medicine.pdf (accessed September 24, 2009).

NCI. 2009b. *The TAILORx breast cancer trial: TAILORx: Testing personalized treatment for breast cancer.* http://www.cancer.gov/clinicaltrials/digestpage/Tailorx (accessed September 16, 2009).

Paik, S., S. Shak, G. Tang, C. Kim, J. Baker, M. Cronin, F. L. Baehner, M. G. Walker, D. Watson, T. Park, W. Hiller, E. R. Fisher, D. L. Wickerham, J. Bryant, and N. Wolmark. 2004. A multigene assay to predict recurrence of tamoxifen-treated, node-negative breast cancer. *New England Journal of Medicine* 351(27):2817–2826.

Paik, S., G. Tang, S. Shak, C. Kim, J. Baker, W. Kim, M. Cronin, F. L. Baehner, D. Watson, J. Bryant, J. P. Costantino, C. E. Geyer, Jr., D. L. Wickerham, and N. Wolmark. 2006. Gene expression and benefit of chemotherapy in women with node-negative, estrogen receptor-positive breast cancer. *Journal of Clinical Oncology* 24(23):3726–3734.

Palomaki, G. E., L. A. Bradley, M. P. Douglas, K. Kolor, and W. D. Dotson. 2009. Can UGT1A1 genotyping reduce morbidity and mortality in patients with metastatic colorectal cancer treated with irinotecan? An evidence-based review. *Genetics in Medicine* 11(1):21–34.

PGxHealth. 2009. *The PGxpredict: Rituximab test: Helping individualize treatment for follicular non-Hodgkin's lymphoma patients.* http://www.pgxhealth.com/rituximab/ (accessed September 21, 2009).

Phillips, K. A. 2008. Closing the evidence gap in the use of emerging testing technologies in clinical practice. *JAMA* 300(21):2542–2544.

Roche. 2008. *Herceptin given prior to surgery improves the chance of survival without relapse for women with HER2-positive breast cancer.* http://www.roche.com/investors/ir_update/ inv-update-2008-12-11.htm (accessed September 16, 2009).

Romond, E. H., E. A. Perez, J. Bryant, V. J. Suman, C. E. Geyer, Jr., N. E. Davidson, E. Tan-Chiu, S. Martino, S. Paik, P. A. Kaufman, S. M. Swain, T. M. Pisansky, L. Fehrenbacher, L. A. Kutteh, V. G. Vogel, D. W. Visscher, G. Yothers, R. B. Jenkins, A. M. Brown, S. R. Dakhil, E. P. Mamounas, W. L. Lingle, P. M. Klein, J. N. Ingle, and N. Wolmark. 2005. Trastuzumab plus adjuvant chemotherapy for operable HER2-positive breast cancer. *New England Journal of Medicine* 353(16):1673–1684.

SACGHS (Secretary's Advisory Committee on Genetics, Health, and Society). 2006. Coverage and Reimbursement of Genetic Tests and Services. Bethesda, MD: SACGHS.

SACGHS. 2008a. *SACGHS—fifteenth meeting—Wednesday, February 13, 2008, Vol. II.* Washington, DC: SACGHS.

SACGHS. 2008b. *U.S. system of oversight of genetic testing: A response to the charge of the Secretary of Health and Human Services.* Bethesda, MD: SACGHS.

SACGT (Secretary's Advisory Committee on Genetic Testing). 2000a. *Enhancing the oversight of genetic tests: Recommendations of the Secretary's Advisory Committee on Genetic Testing.* Bethesda, MD: SACGT.

SACGT. 2000b. *Highlights of the fifth meeting of the Secretary's Advisory Committee on Genetic Testing.* Bethesda, MD: Office of Biotechnology Activities, National Institutes of Health.

Scaltriti, M., and J. Baselga. 2006. The epidermal growth factor receptor pathway: A model for targeted therapy. *Clinical Cancer Research* 12(18):5268–5272.

Schadt, E. E., J. Lamb, X. Yang, J. Zhu, S. Edwards, D. Guhathakurta, S. K. Sieberts, S. Monks, M. Reitman, C. Zhang, P. Y. Lum, A. Leonardson, R. Thieringer, J. M. Metzger, L. Yang, J. Castle, H. Zhu, S. F. Kash, T. A. Drake, A. Sachs, and A. J. Lusis. 2005. An integrative genomics approach to infer causal associations between gene expression and disease. *Nature Genetics* 37(7):710–717.

Schroder, F. H., J. Hugosson, M. J. Roobol, T. L. J. Tammela, S. Ciatto, V. Nelen, M. Kwiatkowski, M. Lujan, H. Lilja, M. Zappa, L. J. Denis, F. Recker, A. Berenguer, L. Maattanen, C. H. Bangma, G. Aus, A. Villers, X. Rebillard, T. van der Kwast, B. G. Blijenberg, S. M. Moss, H. J. de Koning, A. Auvinen, and E. I. the. 2009. Screening and Prostate-Cancer Mortality in a Randomized European Study. *New England Journal of Medicine* 360(13):1320–1328.

Search Collaborative Group, E. Link, S. Parish, J. Armitage, L. Bowman, S. Heath, F. Matsuda, I. Gut, M. Lathrop, and R. Collins. 2008. SLCO1B1 variants and statin-induced myopathy—a genomewide study. *New England Journal of Medicine* 359(8):789–799.

Shak, S. 2009. *Technological hurdles: Diagnostic developer perspective.* PowerPoint presentation. Washington, DC: Genomic Health, Inc.

Smith, B. D., M. Levis, M. Beran, F. Giles, H. Kantarjian, K. Berg, K. M. Murphy, T. Dauses, J. Allebach, and D. Small. 2004. Single-agent CEP-701, a novel FLT3 inhibitor, shows biologic and clinical activity in patients with relapsed or refractory acute myeloid leukemia. *Blood* 103(10):3669–3676.

Sweet-Cordero, A., S. Mukherjee, A. Subramanian, H. You, J. J. Roix, C. Ladd-Acosta, J. Mesirov, T. R. Golub, and T. Jacks. 2005. An oncogenic KRAS2 expression signature identified by cross-species gene-expression analysis. *Nature Genetics* 37(1):48–55.

Teutsch, S. M., L. A. Bradley, G. E. Palomaki, J. E. Haddow, M. Piper, N. Calonge, W. D. Dotson, M. P. Douglas, A. O. Berg, and E. W. Group. 2009. The Evaluation of Genomic Applications in Practice and Prevention (EGAPP) Initiative: Methods of the EGAPP Working Group. *Genetics in Medicine* 11(1):3–14.

van't Veer, L. J., H. Dai, M. J. van de Vijver, Y. D. He, A. A. M. Hart, M. Mao, H. L. Peterse, K. van der Kooy, M. J. Marton, A. T. Witteveen, G. J. Schreiber, R. M. Kerkhoven, C. Roberts, P. S. Linsley, R. Bernards, and S. H. Friend. 2002. Gene expression profiling predicts clinical outcome of breast cancer. *Nature* 415(6871):530–536.

Vose, J., F. Loberiza, J. Armitage, P. Bierman, R. Bociek, and D. Dornan. 2009. Effects of FCGR3A and FCGR2A polymorphisms on outcomes of patients with diffuse large B-cell lymphoma treated with CHOP-like chemotherapy versus CHOP-rituximab. *Journal of Clinical Oncology (Meeting Abstracts)* 27(15S): Abstract No: 8567.

Wolff, A. C., M. E. Hammond, J. N. Schwartz, K. L. Hagerty, D. C. Allred, R. J. Cote, M. Dowsett, P. L. Fitzgibbons, W. M. Hanna, A. Langer, L. M. McShane, S. Paik, M. D. Pegram, E. A. Perez, M. F. Press, A. Rhodes, C. Sturgeon, S. E. Taube, R. Tubbs, G. H. Vance, M. van de Vijver, T. M. Wheeler, D. F. Hayes, American Society of Clinical Oncology, and College of American Pathologists. 2007. American Society of Clinical Oncology/College of American Pathologists guideline recommendations for human epidermal growth factor receptor 2 testing in breast cancer. *Journal of Clinical Oncology* 25(1):118–145.

Woolf, S. H. 2008. The meaning of translational research and why it matters. *JAMA* 299(2):211–213.

Yang, X., J. L. Deignan, H. Qi, J. Zhu, S. Qian, J. Zhong, G. Torosyan, S. Majid, B. Falkard, R. R. Kleinhanz, J. Karlsson, L. W. Castellani, S. Mumick, K. Wang, T. Xie, M. Coon, C. Zhang, D. Estrada-Smith, C. R. Farber, S. S. Wang, A. van Nas, A. Ghazalpour, B. Zhang, D. J. MacNeil, J. R. Lamb, K. M. Dipple, M. L. Reitman, M. Mehrabian, P. Y. Lum, E. E. Schadt, A. J. Lusis, and T. A. Drake. 2009. Validation of candidate causal genes for obesity that affect shared metabolic pathways and networks. *Nature Genetics* 41(4):415–423.

Acronyms

ACCE	Analytic validity; Clinical validity; Clinical utility; and Ethical, legal, and social implications
AHRQ	Agency for Healthcare Research and Quality
AML	acute myelogenous leukemia
ASCO	American Society of Clinical Oncology
CALGB	Cancer and Leukemia Group B
CAP	College of American Pathologists
CDC	Centers for Disease Control and Prevention
CDRH	Center for Devices and Radiological Health
CF	cystic fibrosis
CHOP	cyclophosphamide, doxorubicin, vincristine, prednisone therapy
CLIA	Clinical Laboratory Improvement Amendments
CMS	Centers for Medicare & Medicaid Services
CPT	current procedural terminology
EGAPP	Evaluation of Genomic Applications in Practice and Prevention
EGFR	epidermal growth factor receptor
EMEA	European Medicines Agency
ER	estrogen receptor

FDA	Food and Drug Administration
FISH	fluorescent in situ hybridization
FLT3	FMS-like tyrosine kinase 3
GAPPNet	Genomic Applications in Practice and Prevention Network
HER2	human epidermal growth factor receptor 2
HHS	U.S. Department of Health and Human Services
HRSA	Health Resources and Services Administration
IOM	Institute of Medicine
IVDMIA	In Vitro Diagnostic Multivariate Index Assay
KRAS	v-Ki-ras2 Kirsten rat sarcoma viral oncogene homolog
MEDCAC	Medicare Evidence Development and Coverage Advisory Committee
NCI	National Cancer Institute
NIH	National Institutes of Health
NLA	national limitation amount
OIVD	Office of In Vitro Diagnostic Devices
PCR	polymerase chain reaction
PET	positron emission tomography
PFS	progression-free survival
PMA	premarket application
PSA	prostate-specific antigen
SACGHS	Secretary's Advisory Committee on Genetics, Health, and Society
SACGT	Secretary's Advisory Committee on Genetic Testing
UGT1A1	UDP glucuronosyltransferase 1 family, polypeptide A1
USPSTF	U.S. Preventive Services Task Force

Glossary

Allele – any one of a series of two or more different genes that occupy the same position (locus) on a chromosome.

Analytical validity – the accuracy of a test in detecting the specific entity that it was designed to detect. This accuracy does not imply any clinical significance, such as diagnosis.

Clinical trial – a formal study carried out according to a prospectively defined protocol that is intended to discover or verify the safety and effectiveness of procedures or interventions in humans.

Clinical utility – the clinical and psychological benefits and risks of positive and negative results of a given technique or test.

Clinical validity – the accuracy of a test for a specific clinical purpose, such as diagnosing or predicting risk for a disorder.

Companion Diagnostic Test – in this report, companion diagnostic tests include tests that are predictive of a therapeutic response that have gone through the FDA approval process.

Diagnostic – the investigative tools and techniques used in biological studies or to identify or determine the presence of a disease or other condition.

Epidermal growth factor receptor (EGFR) – a receptor that is over-produced in several solid tumors, including breast and lung cancers. Its overproduction is linked to a poorer prognosis because it enables cell proliferation, migration, and the development of blood vessels. Several new drugs recently approved by the Food and Drug Administration specifically target EGFR.

Genomics – the study of all of the nucleotide sequences, including structural genes, regulatory sequences, and noncoding DNA segments, in the chromosomes of an organism or tissue sample. One example of the application of genomics in oncology is the use of microarray or other techniques to uncover the genetic "fingerprint" of a tissue sample. This genetic fingerprint is the pattern that stems from the variable expression of different genes in normal and cancer tissues.

Herceptin – see Human epidermal growth factor receptor 2.

Human epidermal growth factor receptor 2 (HER2) – a growth factor receptor that is used as a breast cancer biomarker for prognosis and treatment with the drug trastuzumab (Herceptin), which targets the HER2 protein. The HER2 protein is overexpressed in approximately 25 percent of breast cancer patients due to amplification of the gene.

Laboratory-Developed Tests – in this report, laboratory-developed tests include tests that are predictive of a therapeutic response that have not gone through the FDA approval process. The laboratories that provide these tests are, however, subject to oversight by CMS under CLIA.

Off-label use – using a drug that either has not been approved by the Food and Drug Administration or has not been approved for the purpose for which it is being used.

Phase I trial – clinical trial in a small number of patients in which the toxicity and dosing of an intervention are assessed.

Phase II trial – clinical trial in which the safety and preliminary efficacy of an intervention are assessed in patients.

Phase III trial – large-scale clinical trial in which the safety and efficacy of an intervention are assessed in large numbers of patients. The Food and Drug Administration generally requires new drugs to be tested in Phase III trials before they can be put on the market.

Predictive tests – in this report, tests that are predictive of a therapeutic response are referred to as "predictive tests."

Premarket approval – a Food and Drug Administration approval for a new test or device that enables it to be marketed for clinical use. To receive this approval, the manufacturer of the product must submit clinical data showing the product is safe and effective for its intended use.

Premarket notification or 510(k) – a Food and Drug Administration review process that enables a new test or device to be marketed for clinical use. This review process requires manufacturers to submit data showing the accuracy and precision of their product, as well as, in some cases, its analytical sensitivity and specificity. Manufacturers also have to provide documentation supporting the claim that their product is substantially equivalent to one already on the market. This review does not typically consider the clinical safety and effectiveness of the product.

Proficiency testing – laboratories performing non-waived tests must enroll laboratory personnel in tests specific to the subspecialty relevant to the tests they will be evaluating. The Clinical Laboratory Improvement Amendments require proficiency testing of personnel at least once every 2 years.

Proteomics – the study of the structure, function, and interactions of the proteins produced by the genes of a particular cell, tissue, or organism. The application of proteomics in oncology may involve mass spectroscopy, two-dimensional polyacrylamide gel electrophoresis, protein chips, and other techniques to uncover the protein "fingerprint" of a tissue sample. This protein fingerprint is the pattern that stems from the various amounts and types of all the proteins in the sample.

Qualification – the evidentiary process of linking an assay with biological and clinical endpoints that is dependent on the intended application.

***Ras* gene** – a gene encoding for a signal transduction protein that has been found to cause cancer when the gene is altered (mutated). Agents that block its activity may stop the growth of cancer.

Trastuzumab – see Human epidermal growth factor receptor 2.

Validation – the process of assessing the assay or measurement performance characteristics.

Appendix A

Workshop Agenda

National Cancer Policy Forum
Workshop on
Policy Issues in the Development of Personalized Medicine in Oncology

National Academy of Sciences Building – Lecture Room
2100 C Street, NW
Washington, DC 20001

DAY 1: MONDAY, JUNE 8, 2009

8:00 am   **Registration and Continental Breakfast**

8:30 am   **Welcome from National Cancer Policy Forum and**
            Overview of the Workshop
            David Parkinson, Nodality, Inc.
            Roy Herbst, M.D. Anderson Cancer Center

8:45 am   **Technological Hurdles: Vignettes**
            Multiple Genetic Changes in Breast Cancer – *Stephen Friend,*
                Sage Bionetworks
            KRAS in Colorectal Cancer – *Rafael Amado, GlaxoSmithKline*
            EGFR in Lung Cancer – *Bruce Johnson, Dana-Farber Cancer*
                Institute
            FLT3 in Leukemia – *Donald Small, Sidney Kimmel*
                Comprehensive Cancer Center at Johns Hopkins
            Pharmacogenomic Issues in Drug Development – *Mark*
                Ratain, University of Chicago Medical Center
            Moderator – *Fred Appelbaum, Fred Hutchinson Cancer*
                Research Center

10:30 am   **BREAK**

10:45 am   **Technological Hurdles: Professional Perspectives**
           Clinician Perspective – *Richard Schilsky, University of Chicago;*
              Cancer and Leukemia Group B
           Drug Developer Perspective – *Robert Mass, Genentech, Inc.*
           Diagnostic Developer Perspective – *Steven Shak, Genomic*
              Health, Inc.
           Patient Perspective – *Amy Bonoff, National Breast Cancer*
              Coalition
           Moderator – *Richard Schilsky, University of Chicago; Cancer*
              and Leukemia Group B

12:45 pm   **LUNCH**

1:45 pm   **Regulatory Hurdles: Overview of Past Recommendations**
           The Original Secretary's Advisory Committee on Genetic
              Testing – *Wylie Burke, University of Washington*
           The Secretary's Advisory Committee on Genomics, Health and
              Society – *Andrea Ferreira-Gonzalez, Virginia Commonwealth*
              University
           Moderator – *Steven Gutman, University of Central Florida*

2:45 pm   **BREAK**

3:00 pm   **Regulatory Hurdles: What Is the Status Quo?**
           What Is the Food and Drug Administration Currently Doing?
              – *Alberto Gutierrez, Office of In Vitro Diagnostic Devices,*
              Food and Drug Administration
           How Do the Clinical Laboratory Improvement Amendments
              Oversee Laboratory-Developed Tests? – *Penelope Meyers,*
              Division of Laboratory Services, Centers for Medicare &
              Medicaid Services
           What Can the Center for Disease Control and Prevention Do
              to Help in the Assessment of New Tests? – *Ralph Coates,*
              Coordinating Center for Health Promotion, Center for
              Disease Control and Prevention
           Moderator – *Steven Gutman, University of Central Florida*

4:30 pm   **Adjourn Day 1**

DAY 2: TUESDAY, JUNE 9, 2009

8:00 am   **Registration and Continental Breakfast**

8:30 am   **Regulatory Hurdles: Looking Forward**
Why the Food and Drug Administration Should Do More
  – *Robert Mass, Genentech, Inc.*
Is the Status Quo Appropriate? – *Debra Leonard, Weill Cornell Medical College*
Moderator – *Steven Gutman, University of Central Florida*

10:00 am   **BREAK**

10:15 am   **Reimbursement Hurdles**
Medicare Coverage and Reimbursement – *Jeffrey Roche, Coverage and Analysis Group, Centers for Medicare & Medicaid Services* and *Amy Bassano, Center for Medicare Management, Centers for Medicare & Medicaid Services*
Clinician Perspective – *Daniel Hayes, University of Michigan Comprehensive Cancer Center*
Policy Perspective – *Bruce Quinn, Foley Hoag, LLP*
Moderator – *Peter Bach, Memorial Sloan-Kettering Cancer Center*

12:15 pm   **BREAK – Please Retrieve Prepared Lunch and Return for Working Lunch with Last Session**

12:30 pm   **Reactions to the Workshop**
Patient Perspective – *Robert Erwin, Marti Nelson Cancer Foundation*
Industry Perspective – *Stephen Friend, Merck & Co, Inc.* and *David Parkinson, Nodality, Inc.*
Clinician Perspective – *Roy Herbst, M.D. Anderson Cancer Center*
Venture Capital Perspective – *Risa Stack, Kleiner Perkins Caulfield and Byers*
Moderator – *Gail Javitt, Genetics and Public Policy Center, Johns Hopkins University*

1:30 pm   **Adjourn Day 2**

Appendix B

Workshop Speakers and Moderators

Rafael G. Amado, Oncology Medicine Development Center, GlaxoSmithKline

Fred Appelbaum, Clinical Research Division, Fred Hutchinson Cancer Research Center

Peter B. Bach, Memorial Sloan-Kettering Cancer Center

Amy Bassano, Center for Medicare Management, Centers for Medicare & Medicaid Services

Amy Bonoff, National Breast Cancer Coalition

Wylie Burke, Department of Medical History & Ethics, University of Washington

Ralph Coates, Coordinating Center for Health Promotion, Centers for Disease Control and Prevention

Robert Erwin, Marti Nelson Cancer Foundation

Andrea Ferreira-Gonzalez, Department of Pathology, Virginia Commonwealth University

Stephen Friend, Sage Bionetworks

Alberto Gutierrez, Office of In Vitro Diagnostic Devices, Food and Drug Administration

Steven Gutman, College of Medicine, University of Central Florida

Daniel F. Hayes, University of Michigan Comprehensive Cancer Center

Roy S. Herbst, Thoracic/Head & Neck Medical Oncology, M.D. Anderson Cancer Center

Gail Javitt, Genetics and Public Policy Center, Johns Hopkins University

Bruce E. Johnson, Dana-Farber Cancer Institute

Debra G. B. Leonard, Weill Cornell Medical College, New York-Presbyterian Hospital

Robert Mass, Genentech, Inc.

Penelope Meyers, Division of Laboratory Services, Centers for Medicare & Medicaid Services

David R. Parkinson, Nodality, Inc.

Bruce Quinn, Foley Hoag, LLP

Mark J. Ratain, University of Chicago Hospitals

Jeffrey C. Roche, Coverage and Analysis Group, Centers for Medicare & Medicaid Services

Richard Schilsky, University of Chicago, Cancer and Leukemia Group B

Steven Shak, Genomic Health, Inc.

Donald Small, Sidney Kimmel Comprehensive Cancer Center

Risa Stack, Kleiner Perkins Caulfield and Byers